ENVIRONMENT AND MAN
VOLUME TWO

Food, Agriculture and the Environment

ENVIRONMENT AND MAN: VOLUME TWO

Other titles in this Series

Volume 1 *Energy Resources and the Environment*

Volume 3 *Health and the Environment*

Volume 4 *Reclamation*

ENVIRONMENT AND MAN
VOLUME TWO

Food, Agriculture and the Environment

General Editors

John Lenihan

O.B.E., M.Sc., Ph.D., C.Eng., F.I.E.E., F.Inst.P., F.R.S.E.

Director of the Department of Clinical Physics and Bio-Engineering, West of Scotland Health Boards, Professor of Clinical Physics, University of Glasgow, Chairman of the Scottish Technical Education Council.

and

William W Fletcher

B.Sc., Ph.D., F.L.S., F.I.Biol., F.R.S.E.

Professor of Biology and Past Dean of the School of Biological Sciences, University of Strathclyde; Chairman of the Scottish Branch of the Institute of Biology; President of the Botanical Society of Edinburgh.

1976

ACADEMIC PRESS·NEW YORK & SAN FRANCISCO

A Subsidiary of Harcourt Brace Jovanovich, Publishers

Blackie & Son Limited
Bishopbriggs
Glasgow G64 2NZ

450 Edgware Road
London W2 1EG

First published 1976

International Standard Book Number

0-12-443502-5

Library of Congress Catalog Card Number

75-37435

Printed in Great Britain by
Thomson Litho Ltd., East Kilbride, Scotland

Background to Authors

Environment and Man: Volume Two

MALCOLM SLESSER, B.Sc., Ph.D., C.Eng., M.I.Chem.E., F.R.G.S., is Director of the Energy Analysis Unit, University of Strathclyde, and Senior Lecturer in the Department of Pure and Applied Chemistry. He is the author of eight books and is a well-known mountain explorer.

JOY TIVY, B.A., B.Sc., Ph.D., F.R.G.S., is Reader in Geography at the University of Glasgow. She is Honorary Secretary and Editor of the Scottish Field Studies Association, and a member of the Nature Conservancy Council's Advisory Committee for Scotland.

NORMAN W. PIRIE, M.A., F.R.S., was Head of the Department of Biochemistry, Rothamsted Experimental Station from 1947 to 1973. He is a visiting Professor at the University of Reading and at the Indian Statistical Institute, Calcutta.

G. S. HARTLEY, D.Sc.(Lond.), F.N.Z.I.C., M.Inst.Biol., started an academic career in physical chemistry, particularly of colloidal solutions, and was diverted by war-time duties to research in toxic chemicals. After the war, his field of interest changed from humans to pests and weeds, and he was Research Director of Fisons Ltd. (Pest Control Division) before migrating to New Zealand on retirement.

NOEL ROBERTSON, B.Sc., M.A., Ph.D., Dip.Agric.Sc., F.R.S.E., is Professor of Agriculture and Rural Economy in the University of Edinburgh, and Principal of the East of Scotland College of Agriculture. From 1959 to 1968 he was Professor of Botany in the University of Hull.

HUGH PATERSON DONALD, C.B.E., D.Sc., F.R.S.E., is Honorary Professor in the Edinburgh School of Agriculture. He was Director of the Animal Breeding Research Organization from 1950 to 1973, and is now Honorary Scientific Secretary of the Edinburgh Centre of Rural Economy.

Series Foreword

MAN IS A DISCOVERING ANIMAL—SCIENCE IN THE SEVENTEENTH CENTURY, scenery in the nineteenth and now the environment. In the heyday of Victorian technology—indeed until quite recently—the environment was seen as a boundless cornucopia, to be enjoyed, plundered and re-arranged for profit.

Today many thoughtful people see the environment as a limited resource, with conservation as the influence restraining consumption. Some go further, foretelling large-scale starvation and pollution unless we turn back the clock and adopt a simpler way of life.

Extreme views—whether exuberant or gloomy—are more easily propagated, but the middle way, based on reason rather than emotion, is a better guide for future action. This series of books presents an authoritative explanation and discussion of a wide range of problems related to the environment, at a level suitable for practitioners and students in science, engineering, medicine, administration and planning. For the increasing numbers of teachers and students involved in degree and diploma courses in environmental science the series should be particularly useful, and for members of the general public willing to make a modest intellectual effort, it will be found to present a thoroughly readable account of the problems underlying the interactions between man and his environment.

Preface

THIS VOLUME DEALS MAINLY WITH PROBLEMS ARISING FROM THE cultivation of plants and animals for food—the first and still the major activity relating man to his environment. As Dr Slesser explains, food production uses energy in many ways, and the conversion is not always managed economically. Since the natural productivity of the soil, unsupported by machinery, pesticides, irrigation and chemical fertilizers, would not feed a tenth of the world's population, agriculture must make heavy demands on technology. Even so, it is surprising to learn that British agriculture accounts for 10 per cent of the country's energy consumption, and that the calorific value of food that we eat is less than one sixth of the energy (in addition to sunlight) required to produce it.

Dr Tivy examines some of the environmental hazards of intensive cultivation—an important issue, since not much more than 10 per cent of the earth's land surface is suitable for the growing of crops, and most of that is already in use. Alteration of water balance, changes in microclimate, soil erosion and deterioration of soil structure may result from over-enthusiastic cultivation or inadequate husbandry. Here, as elsewhere in the environmental debate, the conflicting needs of short-term economics and long-term ecology have still to be reconciled.

Mr Pirie emphasizes the need for fresh thinking on the protein shortage—the major part of the food shortage which is already apparent in the developing countries and is likely to grow worse. The use of animals to convert plant protein into meat is traditional but wasteful. Plant protein can be made palatable by modern techniques but is largely wasted; leaves, stems, flowers, and seeds have enormous potential as sources of food. New crops and novel methods of farming should also be exploited; the necessary change in food habits need not be unduly difficult.

Technical improvements in agriculture are not always universally applauded. Modern chemical aids to crop production, beginning with DDT, have been criticized—and sometimes banned—because of anxiety over their effects on birds and other wild life. Dr Hartley, reviewing the problems facing producers and users of agricultural chemicals, illuminates the need to educate the public to a better understanding of benefits and hazards. He suggests that present methods of control by legislation and government committees have unexpected side effects, sometimes increasing the dangers that they seek to control.

Finally Professor Robertson and Professor Donald examine the environmental consequences of the domestication of plants and animals in the interests of food production—and of the increasing industrialization of agriculture. Looking to the future, they see British agriculture under pressure to produce more grain at the expense of milk and meat. In the global view, they see the power to survive as resting with those who have the food, and they question the willingness of the well-fed nations to share their meal with the rest.

These five chapters do not exhaust the subject. Other volumes in the series will deal with the chemical environment and the reclamation of barren land. Meanwhile the present volume is offered as an expert appraisal of environmental problems associated with the winning of food—an activity that concerns each of us every hour of the day.

Contents

CHAPTER ONE

ENERGY REQUIREMENTS OF AGRICULTURE

MALCOLM SLESSER

Introduction

Food is a natural product. F. Spencer Chapman (1949) in his book, *The Jungle is Neutral*, showed that the natural fruits, nuts and game of a tropical forest could sustain a small roving band of soldiers indefinitely. Is it just fashion or laziness that has led men to farm with fossil energy? Or is it necessity?

The productivity of a natural eco-system is around 6 kg per hectare* per year of protein suitable for humans, and this is only obtainable from the better land on the earth's surface. Conceivably, such an unintensified system might support around two hundred million people, a figure surpassed by the Middle Ages. Today the world contains almost four thousand million people, with a population doubling time of about 25 years. We have no choice but to use energy-driven technology to raise the productivity of land—a productivity which in some of today's most intensive systems produces fifty times as much protein per hectare as a natural system.

The price we pay for this intensification may be gauged from Table 1.1, which shows the energy consumption for food production in a number of developed countries. Perhaps the most interesting column is that showing energy ratio, calculated by dividing the metabolizable energy made available to the consumer by the heat of combustion of the fuels used to make the inputs. For the United Kingdom and the United States today, that figure is less than one sixth. If we assume an average human food intake of 2500 kilocalories per day, then those of us in the United Kingdom are effectively 'eating' 30 million tons of oil per year. The 220 million people

* 1 hectare (ha) = 2·47 acres.

in the United States use more energy for food provision alone than the 600 million people in India use for all purposes.

This chapter will examine what it is that energy does to the productivity of farming, and the ways in which energy and farming may be correlated so that some intelligent judgments may be made about the future. It must

Table 1.1 Energy consumption in food production within developed countries

		Energy consumption				
Country	*Year*	*to farm gate* (10^{15} J)	*farm-gate to dinner plate* (10^{15} J)	*per year total* (10^{15} J)	*Energy ratio*	*Total as % national energy consumption*
USA	1963	2310	4125	6435	6·4	12
UK	1972	340	300	646	6·5	10
Australia	1965/1969	87	121	208	0·3	10
Israel	1969/1970	33	?	?	1·5	20
Hong Kong	1971	?	?	7	?	26
West Germany	1960	102	386	488	?	12

be remembered, of course, that increased energy alone is not the way to improve a yield. The basic requirement is good husbandry. Good husbandry will employ many different agricultural techniques, according to the circumstances in which the farmer finds himself. But whatever these circumstances, each will require inputs, such as tractors, fertilizers, pesticides or irrigation.

How is energy used?

The rectangle in figure 1.1 represents a field. In the course of a year, the farmer adds a variety of things to that field. These things may be physical, like dung, fertilizers or sprays, or they may be in the form of work-

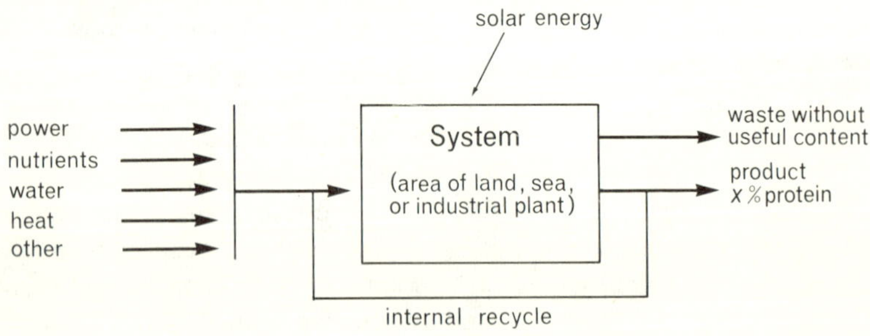

Figure 1.1 Food-producing system.

ploughing, sowing, harrowing, reaping, and so on. Suppose we study the field for a period of one year and break down all the inputs, tangible and intangible, into the amount of input per hectare of the field. Let us assume that the field yields only one product, such as oats or barley or maize. Table 1.2 gives some values for a typical West of Scotland barley field.

Table 1.2 Typical inputs and outputs per hectare of barley production: West of Scotland, 1972 (Turnbull, 1973)

Inputs		*Outputs*	
Herbicide	0·58 kg	Barley	4080 kg (containing 15% moisture)
Nitrogen fertilizer	50 kg		
Phosphorus	32 kg	Straw	2730 kg
Potash	38 kg		
Seed	191 kg		
Tractor time	7 hours		

In Table 1.2, only some of the inputs are readily identifiable in energy terms. Obviously petrol or paraffin is needed for tractors, and we can imagine the large amount of electricity required for the machine-milking of cows, for greenhouse heating or hay drying. But most of the inputs have no obvious energy equivalent. We shall come to these in due course. For the moment we shall look at the direct energy uses. J. R. Stansfield (1974) of the National Institute of Agricultural Engineering, Silsoe, produced a breakdown for UK agriculture and found that the direct energy uses were quite small (Table 1.3).

Table 1.3 Use of direct energy in UK agriculture in 1972 (Stansfield, 1974)

		Tonnes $\times 10^6$	*% UK Energy Consumption*
Oil:	direct use in agriculture	1·58	1·90
	for electricity generation	0·25	0·30
Coal:	for electricity generation	1·00	0·60
	other uses	0·10	0·06

Stansfield found that the oil consumption could be broken down as in Table 1·4.

Table 1.4 Breakdown of oil consumption in UK agriculture, 1972 (Stansfield, 1974)

	Tonnes
Tractors	775,000
Heating, drying, carbon dioxide enrichment	275,000
Greenhouse heating (excluding Channel Isles)	612,000
Road vehicles	82,000
Domestic heating on farms	168,000

Electricity consumption could be broken down as in Table 1.5.

Table 1.5 Breakdown of electricity use in UK agriculture, 1972 (Stansfield, 1974)

	$kWh \times 10^6$
Livestock production	1334
Crop conservation	329
Commercial horticulture	124
Agriculture, general	34
Horticulture, general	41
Domestic (farmhouses)	1315

A great deal of the oil energy is used to run tractors which pull machinery. Stansfield estimated the energy costs of such activities, as shown in Table 1.6.

Table 1.6 Typical fuel consumptions for UK farm tractors in 1972 (Stansfield, 1974)

Operation	*Oil consumption (litres/hectare)*
Ploughing	18
Rotary cultivating	15
Subsoiling	10
Chisel ploughing	9
Disc harrowing	7
Springtime harrowing	6
Drilling, mowing, tedding, baling, distributing fertilizer	3
Rolling	2
Spraying	1

The most surprising thing about Table 1.6 is that so little fuel is used. Pimental (1974) estimated that in Mexico and Guatemala the man-hours required to handle one hectare of land for growing maize, without any mechanical implements, were as high as 1444 (Mexico) and 1415 (Guatemala). There is an obvious attraction in substituting for this considerable effort about 14 imperial gallons of petrol.

However such an analysis would be altogether too superficial. The direct energy inputs do not account for more than half the energy used, in the farming process. Just as we can draw a system diagram for agriculture, like figure 1.1, so we can draw such a diagram for every one of the inputs to the farming process. Energy is required to make the tractor and to maintain it, to make the fertilizers, to synthesize the pesticides, to pump the irrigation water, to make the ploughs, the harrows and the spraying machines. If we are really going to assess the energy in the agricultural process, we have to trace every one of these inputs back, assessing how much energy is needed at each stage. There is no need to trouble the reader with such a tedious operation.

Table 1.7 lists the energy equivalents of a number of standard agricultural inputs. The values are listed as the Gross Energy Requirements (GER), in keeping with the latest recommendations for energy analysis. GER is the actual amount of fossil fuel resource that has to be extracted from the ground in order to make the input available to the consumer, in this case the farmer. It does not include any energy expended in transporting manufactured inputs to the farm.

Table 1.7 Energy equivalents (GER) of agricultural inputs (Leach and Slesser, 1973)

Item	*Unit*	*GER/unit* (10^6 *J/unit*)
Electricity	kWh	14·00
Natural gas	ft^3	1·11
	kg	57·20
diesel oil	imperial gallon	209·00
petrol		192·00
potassium	kg as potassium	9·60
phosphorus	kg as phosphorus	14·00
nitrogen—average	kg as nitrogen	67·00
liquid ammonia	kg as nitrogen	60·00
urea	kg	34·00
	kg as nitrogen	82·00
pesticides and herbicides	average/kg	110·00
ploughing	hectare	836·00
disc harrowing	hectare	325·00
spraying	hectare	45·00

With the data in Table 1.7, Table 1.2 can now be recast in terms of fossil energy extracted from the ground in order to promote the growth of a barley field (Table 1.8). We see that the direct energy is a small proportion of the total energy need.

Table 1.8 Inputs per hectare to a West of Scotland barley field in energy terms (Turnbull, 1973)

Herbicide	50 MJ	
Nitrogen	4880	
Phosphorus	260	1·8 MJ/kg barley
Potassium	310	
Direct fuel	1420	20 MJ/kg protein
	6920 or 6·92 GJ per hectare	

$$6920 \text{ MJ per hectare} = 6{\cdot}92 \text{ GJ per hectare} = 6{\cdot}92 \times 10^9 \text{ J/ha}$$

Energy effectiveness

One suspects that these days a great many inputs to the soil are superfluous. Ehrlich (1970) and others have related stories of unnecessary pesticide treatment, particularly in the cotton crop. Heichel (1973) of the Connecticut Agricultural Experiment Station in the United States has made a study of energy effectiveness. He found that yields of digestible energy from a vegetable garden in New Guinea or from rice cultivation amounted to as much as 14 megacalories of digestible energy per hectare, a figure only slightly bettered by the more technologically advanced farmers of Pennsylvania and Iowa in 1915. He relates that the 'modern oat farmer in

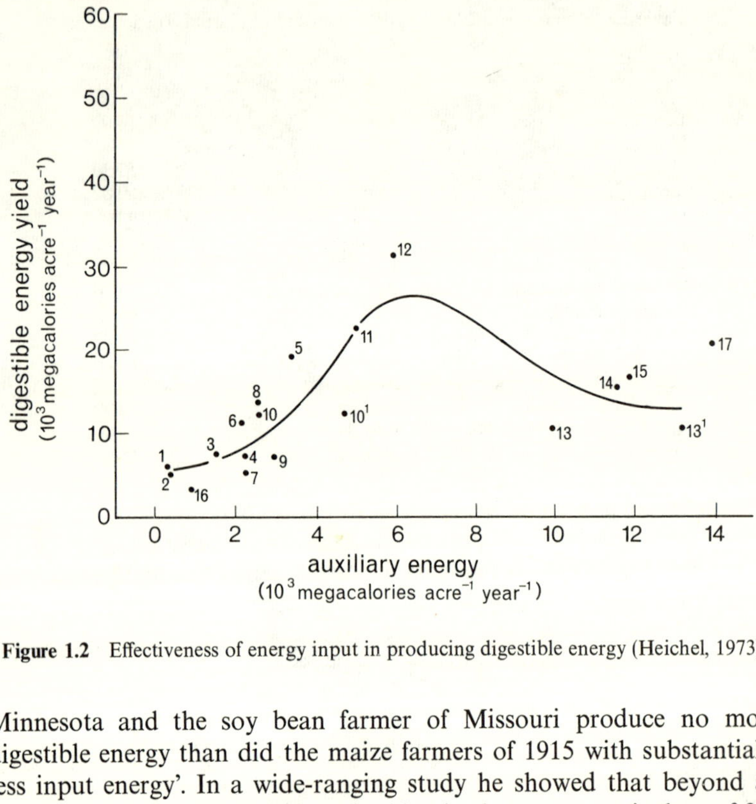

Figure 1.2 Effectiveness of energy input in producing digestible energy (Heichel, 1973).

Minnesota and the soy bean farmer of Missouri produce no more digestible energy than did the maize farmers of 1915 with substantially less input energy'. In a wide-ranging study he showed that beyond an energy input equivalent of 60 GJ/ha (that is, the energy equivalent of 300 imperial gallons of refined oil per hectare) the yield of digestible energy actually declines. Figure 1.2 gives Heichel's plot. The numbers on the plot are identified in Table 1.9.

In figure 1.3 these results are recast in a most interesting form. We see that as the inputs rise, the relative advantage diminishes—the law of diminishing returns.

Heichel's results must be treated with some reservation. They have been deduced using money values for inputs and converting them on a standard basis of the 1970 US energy consumption per dollar of GNP in that year. As the subsequent work of Hannon and Herenden at the Centre for Advanced Computation has shown, the energy used per unit of production value varies widely.

Table 1.9 Number code for crops and cropping systems in relation to figure 1.2

Crop	*Area*	*Date*	*Code*
LEVEL I			
Vegetable	New Guinea	1962	2
LEVEL II			
Irrigated rice	Philippines	1970	1
LEVEL III			
Corn for grain	Iowa	1915	3
Corn for grain	Pennsylvania	1915	4
Corn for silage	Iowa	1915	5
LEVEL IV			
Alfalfa—hay	Missouri	1970	6
Oats	Minnesota	1970	7
Sorghum for grain	Kansas	1970	8
Soybeans	Missouri	1970	9
Sugarcane (without processing)	Hawaii	1970	10
Sugarcane (including processing)	Hawaii	1970	10′
Corn for grain	Illinois	1969	11
Corn for silage	Iowa	1969	12
Sugarbeets (without processing)	California	1970	13
Sugarbeets (including processing)	California	1970	13′
Peanuts	North Carolina	1970	14
Irrigated rice	Louisiana	1970	15
Winter wheat	Montana	1970	16
Potatoes	Maine	1968	17

Heichel's pointer that farming has been tending to use more and more energy for comparatively little gain was well proved by the study by Pimental and his associates at Cornell University, USA. They studied US maize production from 1945 to 1970, giving an energy value for all inputs, including labour.* Their data are summarized in Table 1.10. The energy values shown are not the Gross Energy Requirements (GER) because they have not been traced back to the energy in the ground, but they are unlikely to be more than 10–15% low. The final row of data is modified by 15% to take into account this difference, but even so, the

* The value Pimetal and his associates use is the metabolizable calories consumed, not the energy to sustain the farm worker.

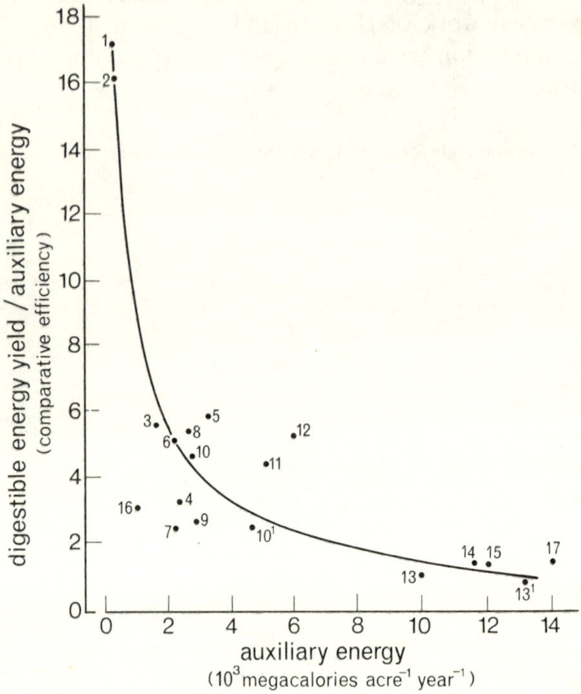

Figure 1.3 Recast of figure 1.2 (1 Mcal acre^{-1} = 10·3 MJ ha^{-1}).

modified figures should be treated with caution due to the rough nature of this correction.

The last two rows of Table 1.10 illustrate the diminishing return from intensification of maize production. Yet something of a plateau was gained in the period 1950 to 1964, possibly due to better technology, better seed and other genetic or technological improvements.

Pimental draws some gloomy conclusions from his study. He points out that if western farming technology plus a US living style were to be exported to the less developed parts of the world, 1200 litres of fuel would be needed for each man, woman and child per year for food supply alone. He reckons this represents about 14 years' consumption of total likely petroleum reserves. Before the end of this chapter we shall return to this topic.

Table 1.10 gives a most useful breakdown of the energy-demanding sectors. Direct fuel, as liquid fuels and electricity, was 62% in 1945, fell sharply and rose to 54% by 1970. Nitrogen has clearly taken the lion's share of the increase. But we should note that the energy for making machinery, plus the fuel to run that machinery, exceeds the energy requirements of the fertilizers.

Table 1.10 Energy inputs per hectare in maize production in the USA (all figures in GJ) (revised after Pimental *et al.*, 1973)

Inputs	*1945*	*1950*	*1954*	*1959*	*1964*	*1970*
Labour	0·30	0·24	0·23	0·2	0·15	0·12
Machinery	4·5	6·2	7·5	8·7	10·5	10·5
Fuel	14·0	15·5	17·2	18·0	18·9	19·9
Nitrogen	1·4	3·0	5·3	8·1	11·6	22·2
Phosphorus	0·25	0·36	0·42	0·58	0·64	1·12
Potassium	0·13	0·25	0·45	0·75	1·02	1·49
Seeds for planting	0·77	0·92	1·13	1·34	1·49	1·49
Irrigation	1·04	1·3	1·50	1·71	1·89	1·89
Insecticides	0	0·03	0·08	0·19	0·28	0·28
Herbicides	0	0·1	0·03	0·07	0·1	0·28
Drying	0·1	0·35	0·74	1·65	2·48	2·98
Electricity	0·8	1·35	2·5	3·47	5·02	7·67
Transportation	0·5	0·75	1·15	1·50	1·74	1·74
Total inputs	23·9	30·8	39·1	47·5	57·2	73·4
Corn yield (output)	77·5	86·65	93·5	123·1	155·0	184·7
GJ return/GJ input	3·24	2·81	2·39	2·59	2·71	2·52
Modified 15% to GER	2·82	2·44	2·07	2·21	2·35	2·19

What we are witnessing in these figures is the industrialization of farming. Clearly it must be having an effect on many aspects of the environment. Labour is being withdrawn from the land, and while the agrichemical business is flourishing, the heavy applications of fertilizer are beginning to cause pollution.

Labour and the land

One of the principal changes in agriculture with the advent of energy-driven inputs is the massive substitution of energy for labour. The Pimental data in Table 1.10 show the energy equivalent of labour falling from 1·2% of the total input in 1945 to 0·16% twenty-five years later, yet the productivity of the land has increased 240%. When economists and farmers talk about the fantastic increase in the productivity of farming, they mean that for each unit of man power the yield has increased enormously—by 600% in this case. Seen in this light, farming has indeed shown remarkable improvement. The Steinharts, who made a time series study of US farming in 1974 estimated that the US farming system is now such that one farmer feeds fifty people. They correlate the man-hours in farming with energy inputs to farming, as in figure 1.4, taken from their paper. It is interesting to note that in the mid-fifties, the peak of farming revolution in the United States, the marginal productivity of labour was $3{\cdot}4 \times 10^{-9}$ man-hours per joule of fossil energy, or it took just over a gallon of petrol to substitute for an hour of a farm-worker's time. By 1970 this

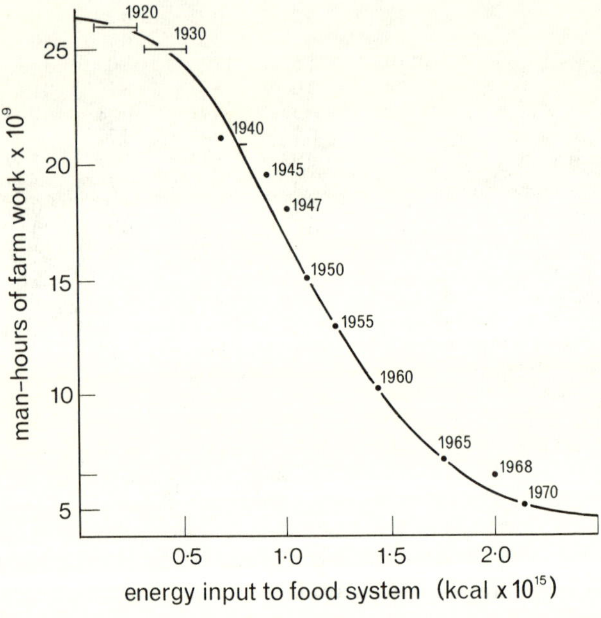

Figure 1.4 Correlation of man-hours of farmwork with energy inputs to farming.

had fallen to $1{\cdot}1 \times 10^{-9}$ man-hours per joule, or 4·3 gallons of petrol for an hour of farm-worker's time. In terms of the energy crisis of 1973 this makes further substitution a most expensive business.

Both Leach (1975) and Blaxter (1974) point to similar changes in the United Kingdom, though the analyses are not yet so detailed. Man-power has fallen to 40% of the 1930 figure, about the same as in the United States, while energy use has quadrupled. Yet it has not all been gain. Leach finds that for every man substituted on the farm, half a man's work is created in the agri-business, while several times that is employed in the food-processing industries and delivery services. The change of system has created many jobs, though many of them may be quite unnecessary, even to our continued well-being.

Energy required to make foods

The statement is often made that we are now 'eating' oil. How true is this in fact? Table 1.11 gives energy requirements for many foods from many lands. Of course, no value is absolute. Different soils, different or indifferent husbandry, difference in seasons, can all make changes. But they are indications of the reality of the above statement. And it seems that when we go for meat and fish, we are indeed eating oil, many times over.

Table 1.11 Energy requirements for typical foods

FOOD	*Country of origin*	*MJ/kg*	*Equivalent kg petrol per kg food*
Vegetables			
Rice	India	3	0·05
	Korea	1·8	0·035
	USA	1·7	0·035
	Japan	1·3	0·002
Wheat	USA	0·9–6	0·014–0·1
Maize	Korea	2·3	0·035
	USA	2·0–6·0	0·035–0·1
Soya		1·6–10	0·035–0·175
Barley	Scotland	0·8	0·018
Hay	Scotland	0·75	0·014
Silage	Scotland	0·5	0·01
Sugar Beet	Scotland	0·3	0·007
Potato	England	0·13	0·0035
	USA	1·6	0·035
Peanuts	USA	1·6–10	0·035–0·175
Animal Protein			
Beef, grass-fed	England	20	0·37
	New Zealand	5	0·1
intensive	USA	48–70	0·8–1·3
Fish, fish farm		26–50	0·5–1·0
Lamb	Scotland	5	0·1
intensive	USA	66	1·2
Milk maximum production		3·0	0·05
Industrial			
Single-cell protein from oil		100	1·9
Yeast from molasses		27	0·5
Lysine synthesis		90	1·7

Significance of energy requirements for food

While Table 1.11 shows us how much energy is required to make foods, it also shows that the amount needed differs from one place to another. Why is this? Is it simply a matter of the soils, the climate, or is there something else? The answer lies in the wide variety of agricultural systems found throughout the world, some of which are much more developed than others, using correspondingly more input energy. We can get a feel for these different agricultural systems if we express the input not just as so much energy per unit of product, but as the amount of energy spent on food growing per unit of land.

The calculation is really quite simple. Using the methods already discussed, we assess all the inputs and convert them to their equivalent to get the Gross Energy Requirement (GER) of the food. We then express this per hectare of land required for the production process. This is called *energy density*. In doing this we must be very careful to include all the land, just as we had to be careful to include all the energy-consuming steps back along the chain of production. For example, if we are assessing the process of making beef by means of a feed-lot, then the total energy requirements are the sum of the energy needed to make all the inputs, plus that needed to run the feed-lot. The agricultural land area required is the total area of land needed to make all the inputs. Thus the UK farmer who intensively breeds beef cattle, not only uses the comparatively small area of his own farm, but also the many acres of farms at home and overseas where are grown the molasses and barley, soya meal and linseed cake with which he fattens his cattle for the market. He is effectively farming far more acres than only his own farm.

Now we can express the yields from the farming enterprise not as so many tons per acre or money profit per man-hour, but as kilograms of protein contained in the produce per hectare; and we find that data from all over the world come together on one plot. There seems to be a very good correlation between the amount of protein man can win from a given area of land and what he puts into that land. The sort of plot that emerges is shown in figure 1.5. What it seems to say is that the more energy we put in, the more protein we get out. Go on putting stuff into the ground and, lo and behold, more food will emerge! This is in fact not the case, and is not how this diagram should be interpreted.

In figure 1.6, using the same coordinates as in figure 1.5, the general relationship between production and inputs is shown as a line. Imagine that we find a piece of agricultural terrain which is not farmed intensively, where, let us say, wheat is grown. The yield for a given input may be given by the lower point of the dotted line. Now, using good husbandry and such ways of increasing wheat yields as are recommended by competent agronomists or agricultural experts, experiments are carried out to see how the yields improve with added inputs. At the same time, the inputs are carefully measured and expressed as energy equivalents. The development of production with inputs will be shown by a plot like the dotted line. Note that eventually production per hectare levels off and then eventually

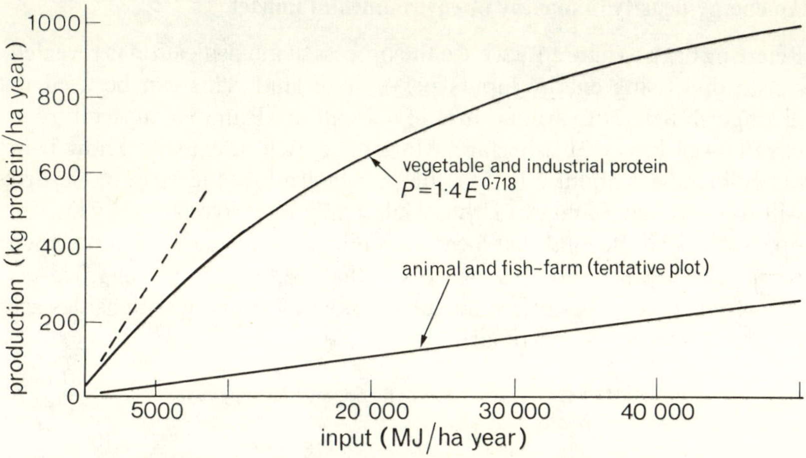

Figure 1.5 Correlation between the amount of protein that can be won from a given area of land with what is put into it.

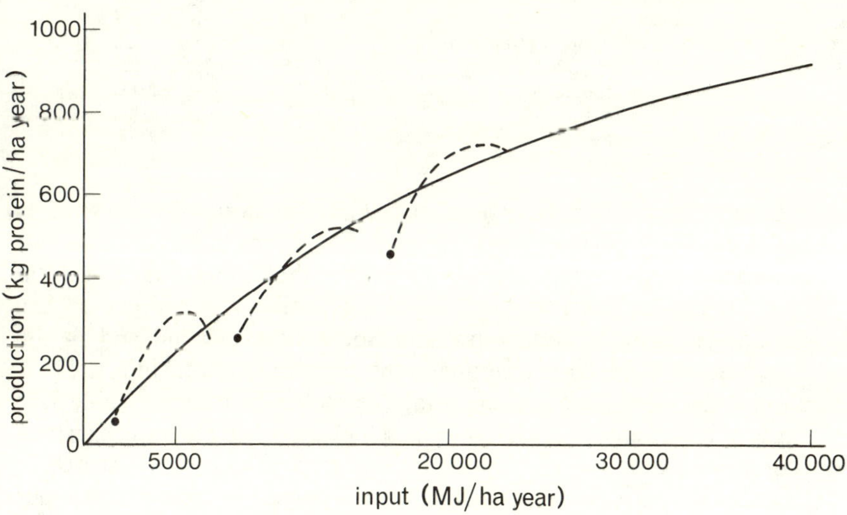

Figure 1.6 Response of individual crops to additional inputs.

falls. That land and that crop have reached their limits. Further development needs another crop or a whole new range of inputs. Perhaps in the end the limiting input may prove to be water, or sunshine, or some trace mineral. Who can say? What we must not assume is that it is possible for any given crop to proceed ineluctably up the line of production versus input.

An energy-density taxonomy of environmental impact

Referring back to figure 1.5 we see the process of intensification is revealed by steadily rising energy inputs per unit of land. This can be used to distinguish between various sorts of agriculture. Primitive agriculture is usually well below 2 GJ/hectare. Marginal agriculture, as we know it in our hill lands, is about 2 to 5 GJ/ha. A modern lowland farm in Europe will use between 5 and 20 GJ/ha, while really intensive farming can run up to 40 GJ/ha. Beyond that farming is into what might be better termed agri-business and employs processes that can only be regarded as industrial. Table 1.12 gives some idea of how these energy intensities are related to actual systems of farming.

Table 1.12 An agricultural taxonomy in energy terms

Examples	*Energy density (GJ/ha)*	*Protein yields (kg/ha)*
hunter-gather	zero	
Andean village (Peru)	0·2	0·5
hill-sheep farming (Scotland)	0·6	1–1·5
marginal farming	4	9
open-range beef farming (New Zealand)	5	130
mixed farm in developed country	12–15	500
intensive crop production	15–20	2,000
feed-lot animal production	40	300
algae manufacture	1600	22,000

In recent times there have been complaints that farming, because of its high intensity, is beginning to cause pollution. Many intensive cattle and pig farmers produce more animal manure than they know what to do with, and this can pollute rivers. Intensive application of fertilizers, as has become common in certain parts of South East England and in the American Mid-West, are resulting in the leaching of nitrate into streams, affecting growths there and upsetting the ecological balance. Table 1.12 suggests that energy density may be a useful yardstick for measuring where farming begins to have an environmental impact. Certainly research in this area is needed, but it looks as if once farming has exceeded an input of 15 gigajoules/hectare, the farmer is no longer keeping his messes to himself, but is sharing them with the neighbourhood.

Food policy planning

The message of energy and agriculture is that where there is energy, so shall we have not only food, but good food. If we can turn fossil fuels into petro-protein, then we can always feed that to the cows and eat beef.

But as we have already quoted from Pimental, there is simply not enough oil for us to do that. Of course, one day far into the future, we may have unlimited fusion energy, and the long-dreamed energy-bountiful world may actually come about. There is no harm in dreaming, but it would be exceptionally unwise to base our plans for the future on an energy source for which so much research and such immense development is still needed. At the best it may be something our grandchildren will enjoy.

In the real world of today, a world in which oil energy may go through a peak rate of consumption by 1995, we have to look at energy and agriculture in very serious terms.

We can gain some perspective if the data in figure 1.5 are recast to ask the question: For a given level of diet, and a given amount of agricultural land per person in a country, how much energy must that country devote to its agricultural sector in order to be self-sufficient?

Figure 1.7 shows two plots. The first is for what we might call a 'survival diet'—vegetable or industrial protein only at 60 g/day/person. The second we might call a 'European diet', containing two-thirds of the necessary protein in the form of fish or meat. What emerges is extremely instructive. The more land there is per person in a country, the less is the amount of energy that *must* be invested and spent in order to secure self-sufficiency in food. As population rises, as land is eroded or taken up for building and roads, the available hectares per person decline, and the minimum energy for food production inescapably rises. It is a rise that technology cannot get round. It is, of course, all implicit in the Laws of Thermodynamics, of which the First Law can be freely translated as there being no such thing as a free meal, and the Second Law, that we cannot break even anyway.

The plot in figure 1.7 showing the European level of diet is fascinating. It carries a deadly message for heavily populated countries. Beyond about 0·1 hectare of land per person, the energy requirements for self-sufficiency are so enormous as to make a 'European level' of diet increasingly and prohibitively expensive.

In Table 1.13 are computed the energy requirements for self-sufficiency for a number of countries. For vegetable diets the energy requirements are not great, though in the case of countries in the same state of development as India, such energy forms a significant percentage of the national energy budget. But the energy requirements for high-quality 'European' diets are very much larger. If we read this table alongside Table 1.1, it can be seen there that in developed countries most of the energy is actually spent between the farm gate and the dinner plate. It follows that the energy requirements for the more densely populated countries could be enormous. Japan is a case in point. Indeed, the traveller to Japan (0·067 ha/capita) is

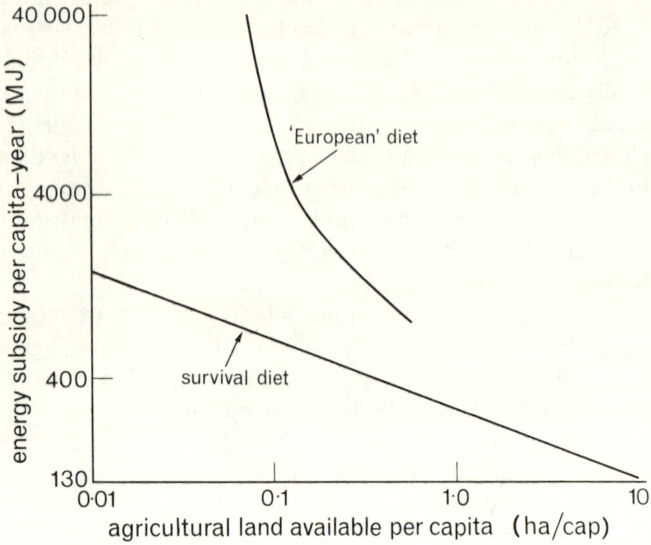

Figure 1.7 Relationship between the energy investment in the land and the land area available.

acutely conscious of the high cost of good protein, whereas in a spacious country such as Argentina, beef is cheap.

The message is this: security in the future is consonant with having either enough space for the promotion of reasonably low-energy agriculture or assured access to ample energy supplies.

Table 1.13 Minimum energy for self-sufficiency in food

Country	*1970 population* $\times 10^6$	*Area of agricultural land per capita (hectares)*	*t.c.e.*[a] *per capita vegetable diet*	*Total energy for vegetable diet (t.c.e.* $\times 10^6$*)*	*Percentage of National 1967 energy consumption*	*Total energy for high-quality diet (t.c.e.* $\times 10^6$*)*
India	572	0·34	0·0145	8·3	13·2	28
China (theory)	730	0·30	0·016	11·7	?	42
China (actual)				30·0	?	
USA (theory)	202	2·24	0·007	1·41	0·05	5·2
USA (actual)		0·76	—	—	1·15	34·6
Korea	30	0·078	0·027	0·81	6·4	9·0
UAR	31	0·093	0·025	0·78	7·5	7·1
Singapore	2·0	0·006	0·10	0·2	negligible	very large
Netherlands	12·7	0·18	0·019	0·24	1·0	1·3
England	48	0·269	0·016	0·77	0·33	3·4
Japan	100	0·067	0·028	2·8	1·5	50·0

[a] t.c.e. is tons coal equivalent = $7{\cdot}00 \times 10^6$ kcal (hard coal).

Ghost acreages

A good many highly developed countries, responding naturally to the dictates of figure 1.7, solve their problems by trade. Japan, Netherlands, Britain and Belgium are prime examples. Food is imported from areas where the intensity, and thus correspondingly the costs, are lower. Within the last two years farmers in Britain were actually encouraged to intensify their beef production, which of course they had to do by importing more feed grains and fish meal. Such a policy could only have been advocated by a government whose advisors saw the economic system as some Pandora's box in which limits did not exist. Farmers in Britain have learnt the sorry truth the hard way, as in the winter of 1974/75 beef prices fell, cattle starved, and some farmers became bankrupt. The reason is clear. These developed countries were living off what George Borgstrom, the famous agronomist from Michigan, calls Ghost Acreages. These are acres beyond the country of consumption. Yet, as the world has filled up with people, as fishing has gone from technological triumph to total vacuum cleaning of the seas, it was inevitable that the available surplus from the Ghost Acreages countries must fall. And when the 1974 energy price rises of 300% came through, the whole pattern of food production in these former areas of surplus changed dramatically.

In Table 1.14 we see that the land in the Netherlands provides a mere 20% of its country's needs. In Japan, it provides only 13%, whereas France

Table 1.14 Ghost acreage of selected countries (all figures in 10^6 hectares) (Borgstrom, 1969)

	Real acreage	*Ghost acreage*		
	Tilled land	*Fish*	*Trade*	*Total*
China	13·7	4·4	0·5	18·6
India	31·7	1·5	1·2	33·4
Indonesia	15·9	3·0	0·7	19·6
Japan	5·8	22·3	15·9	44·0
UK	13·5	9·1	29·1	51·7
France	40·5	4·2	1·3	46·0
Netherlands	7·3	18·1	11·1	36·5

—with a large amount of land per capita—manages to provide 88% of her own needs. But the huge dependence on imported food is clearly shown in the case of the United Kingdom, whose trade acreage is 56%, i.e. of the 0·52 hectare of land that is needed to feed each person, only 0·135 is available in Britain, 0·09 is obtained from fishing, and the rest comes from land overseas. It does not take a vivid imagination to consider what will happen as the population of the food-exporting nations increases, and they begin to cut down on exports. It is already beginning to happen, for a very good economic reason. Many developing nations are finding it more profitable to keep their food at home and not to sell it as a cash crop to buy manufactured goods. Rather they try to make more manufactured goods in their own lands.

What will happen, then, as the ghost acreages available to food-importing countries diminish? The first thing, which is already happening, is that nations will try to sequester larger parts of the oceans for their own use. Thus we have the pressure for a 200-mile zone around our shores. After that, the only recourse is to intensify the land or reduce the quality of the diet. In dietetic terms this would not be serious. Most of us in developed lands eat much more protein than we need, and we could remain healthy with less animal or fish protein. But we like it. It represents part of the quality of the life in a developed country. It is under these circumstances that Table 1.13 takes on a real significance.

The energy-free farm

The natural reaction of many people to the industrialization of our society has resulted in a strong environmental movement, part of which is reflected in the desire of some to get away from factory farming, the prodigious use of inorganic fertilizers, and so on. Some even want to go back to the simple agricultural style of total self-sufficiency.

It must be said right away that a return to the simple agricultural style is not possible, at least for the majority. There is not enough land available to feed the population we have; it is certainly not enough for individual self-sufficient units. Of course, under our land tenure system, the owner of a piece of land can, if he chooses, farm at any level of intensity that pleases him; though, if he farms too intensively, he may run foul of pollution laws. So, if we have wealth enough and do not look upon land as something from which to wrest the maximum cash return, we can farm it unintensively if we choose. Certainly the low-intensity farmers who grew their own winter keep in 1974 were better placed to face the financial pressures of the world of 1975 than the farmers of heavily intensified land with large capital investment, who had to import feed to keep alive a large stock of animals. But the fact is that if every farmer used his land only to keep himself and his family alive, the rest of us would be short of food. Nowadays we might even question whether an individual has the right to get less than the best out of his land. However, over-intensification seems just as great an evil as inadequate use of land.

The best use of land is a hard phrase to define. There are legitimate reasons for abandoning many of the newer highly intensive methods. The yield of crops farmed by a process in which animal manures are recycled can be as large and sometimes larger than that from the use of inorganic fertilizers. Crop rotation, which was once the great farming discovery of Europe, has recently been rediscovered in North America, where monoculture had begun to take sway.

Recycling and lower intensity reduce energy use, and that is good. It may or may not reduce output. But there is plenty of land in most developed countries that is far from properly used, and it seems that modern methods are actually forcing small parcels of land out of use. Is this sensible?

A useful computer study of the effect of changing patterns of farming was made in 1970 by the Iowa State University Agricultural Advisory Centre. They ran a linear-programming model of the whole of the United States, exploring what changes in land use would result from reductions in fertilizer input. They found that if *no inorganic* fertilizers were used at all in the United States, total production of food would fall by only 11%, and that prices would rise 10%.

Conclusion

What conclusions can be drawn about the future, simply from the standpoint of energy and agriculture? The first is that energy and food are related. A lack of the former will soon lead to a lack of the latter. The amount of land available for farming is diminishing all the time, and the population is rising. Land *per capita* is falling fast. Food can only become more expensive in relative terms. The choices before us depend on whether we live in a developing region or in a country which has already been developed.

The developing countries have greater leeway, provided that they can hold down their population growth. For them, as for the developed countries, the attraction of land reclamation and recovery is great—it keeps them at the low end of figure 1.5. The developed nations, on the other hand, face a much more difficult situation. Animal protein must rise in price, relative to other foods, and may, as in Japan today, become a luxury item. The solution in the developed countries cannot lie in buying food from overseas. It has to lie in home-produced food. Consider Scotland, for example: millions of acres of hill land are potentially upgradeable, and those lovely upland pastures could yield grass-fed mutton, beef and venison in an abundance two or three times greater than today.

But all this must have impact upon the environment. The problem can no longer be tackled piecemeal, just as it can no longer be tackled as an exercise in economics.

FURTHER READING

K. Blaxter (1974), *New Scientist*, 14 February, pp. 400–403.

G. Borgstrom (1969), *Too Many*, Collier-McMillan, London, p. 34.

F. S. Chapman (1968), *The Jungle is Neutral*, F. A. Thorpe.

P. and A. Ehrlich (1970), *Population Resources and Environment*, W. H. Freeman, San Francisco.

Energy Analysis, Conventions and Methodology, IFIAS workshop report, International Federation of Institutes of Advanced Study, Nobel House, Sturegatan 14, Box 5344, Stockholm, Sweden, 1974.

G. Heichel (1973), Report of Connecticut Agricultural Experimental Station, Box 1106. New Haven, Connecticut, 06504, USA.

R. Herenden (1963), *Energy Input–Output Matrix for the USA*, CAC Document 69, Centre for Advanced Computation, Urbana, Illinois 61801, USA.

Iowa State University Extension Service, Report Pm-409, 1968, Des Moines, Iowa, USA.

G. Leach (1975), Chapter in *Man Food Equation*, Academic Press, London.

G. Leach and M. Slesser (1973), *Energy Equivalents of Network Inputs to Agriculture*, Strathclyde University.

D. Pimental (1973), *Science*, **182**, p. 443.

D. Pimental (1974), *Proceedings XXV Pan-American Conference Series, Guatemala, 1974*, Instituto de Nutricion de Centro America y Panama, Guatemala City.

J. and S. Seymour (1973), *Self-Sufficiency*, Faber and Faber, London.

J. R. Stansfield (1974), National Institute of Agricultural Engineering, Report 13, West Park, Silsoe, Bedford, England.

J. S. and C. E. Steinhart (1974), *Science*, **184**, pp. 307–316.

G. Turnbull (1973), B.Sc. Thesis. Dept. of Applied Chemistry, University of Strathclyde.

CHAPTER TWO

ENVIRONMENTAL IMPACT OF CULTIVATION

JOY TIVY

Introduction

Fire and the stone axe were the two most important tools with which man began to assert his ecological dominance and to modify drastically his physical environment. They initiated his graduation from a mere food gatherer, completely at the mercy of nature, to a food planter and/or herdsman capable of exploiting and manipulating available organic resources for his own particular and increasingly sophisticated needs. The way in which man has used and continues to use these basic resources, and the intensity of the resultant physical impact on the soil and atmosphere, are dependent on a complex inter-relationship between the nature of the environment, on the one hand, and the particular stage in his cultural and technological development, on the other.

Today, approximately 10% of the earth's land surface is used for the production of crops for food, animal fodder, and various industrial purposes. Most are annual or biennial arable field crops, or sown (improved) perennial grassland; the proportion occupied by the perennial bush and tree crops of orchard or tropical plantations is, in comparison, relatively small. Cropland is concentrated in two main types of area (figure 2.1): humid temperature latitudes, particularly in the extensive lowlands and plains of North America and Eurasia, and the great flood-plains and deltas of the Far East where there are tropical climates with abundant but markedly seasonal or monsoonal rainfall. The land presently under crops does not fall far short of that potentially cultivable, given existing technical knowledge. A very much greater area is, because of unfavourable climate, soil and slope, beyond the practicable limits of cultivation. Of this, some 26% of the total land area is used for the largely uncontrolled grazing of domestic animals. Much is in arid and semi-arid areas, as in the savanna lands of the tropics and the open-range lands of North America; some is 'upland' rough grazing land, as in the case of Mediterranean *maquis* and *garigue*, and British moorlands. On all these range lands, use of the existing vegetation cover for browsing and grazing by domestic herbivores has long been accompanied by the use of fire. Burning to clear away the old (or dead) and useless vegetation to promote greater productivity of more

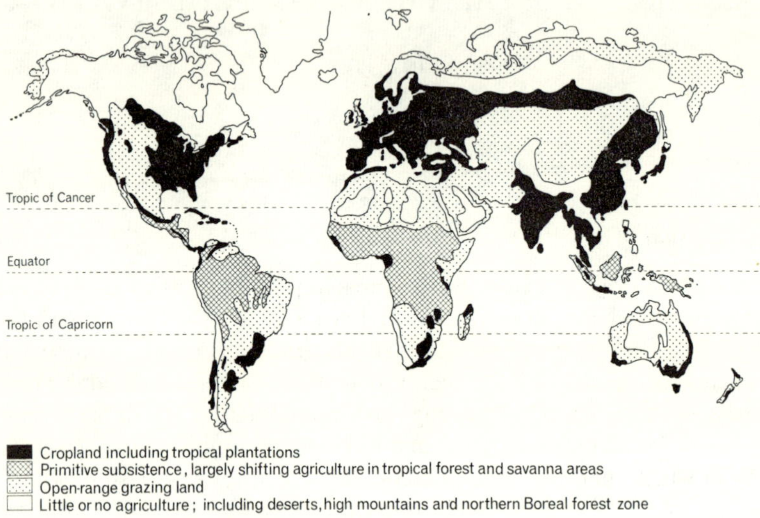

Figure 2.1 Main agricultural regions of the world.

palatable and nutritive forage is a time-honoured method of land management. Forest or woodland, which may once have covered at least two-thirds of the earth's surface, has been reduced to one third of its original area. Even within the two great remaining blocks of the Tropical Rain Forest and the Boreal Forest, the proportion which is in a still relatively unmodified primeval or virgin state is small and rapidly dwindling.

This land use pattern has been produced at the expense of either the complete or partial removal of a former vegetation cover, or a progressive and profound change in its composition, consequent upon selective grazing and burning. Clearance or modification has inevitably been, and still continues to be, accompanied by environmental repercussions. The extent of the latter, however, is dependent on the nature of the pre-existing vegetation and soil cover, and on past and present climatic conditions, together with the degree of modification and the length of time since it was first effected. Among those which have triggered chain reactions which man is still endeavouring to control are:

(*a*) alterations in the water-balance or hydrological cycle; and
(*b*) micro-climatic changes.

The hydrological cycle

The hydrological cycle refers to the global circulation of fresh water, by evaporation, from its main reservoir—the oceans—into the atmosphere. From here it is returned by condensation and precipitation in the form of either dew, mist, rain, snow or hail (directly or from the land) back into the oceans (figure 2.2). The energy required for this process is derived either

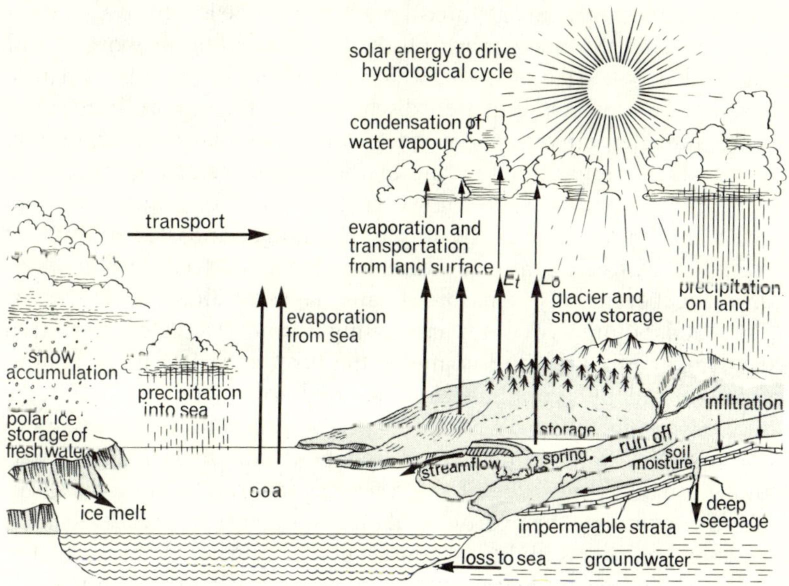

Figure 2.2 The hydrological cycle. About 90% of the precipitation over the land is derived from the sea and only 10% comes from evaporation and transpiration from the land surface (from Pereira, 1973, *Land Use and Water Resources*, Cambridge University Press).

directly from insolation (approximately 45% of the total incoming solar energy is expended in evapo-transpiration) or indirectly as long-wave heat radiation from the surface of the earth. The latter is one of the major factors controlling atmospheric circulation and thus the distribution of precipitation. Hence the hydrological cycle is important for man because it controls the amount of *fresh water available* at any one place at a given time. Of the water input from all forms of precipitation, only about 25% falls on land surfaces and, of this potential supply, the amount and form in which it becomes available to man is determined by the routes by which it is returned to the atmosphere (see figure 2.2). The significance of these routes varies according to his needs and demands for water. The farmer's priorities will differ from those of the industrialist or urban householder.

The soil is the main source of water for all but a very small proportion of land plants. The availability of water for absorption by a plant's roots depends on the depth of the root zone (which can vary from a few centimetres to as much as 15 metres) and the ability of the soil to hold or retain precipitation. Practically all the water taken up by plants is returned to the atmosphere by the process of evapo-transpiration.

However, of the total incoming precipitation, some is intercepted by the vegetation cover and held or 'stored' long enough to be lost by evaporation before it reaches the land surface. Of the latter, a certain proportion will be evacuated by surface run-off, which eventually drains into streams, rivers, lakes, etc., which ultimately discharge into the oceans. Infiltration of water into the soil will be dependent on the amount of precipitation, less that which is intercepted or runs off the surface. Return of water from the soil may be by two routes however: (1) by way of evaporation and evapo-transpiration, and (2) by way of the longer and slower route of percolation, as freely draining soil-water, down to ground-water levels from where it eventually seeps into streams, lakes, or directly into the sea.

The total volume of water circulating through a given terrestrial ecosystem, and its distribution among the possible routes it can follow, are determined by prevailing climatic conditions, landforms, and the nature of the vegetation cover and the underlying soil. A shift in any of these variables will be reflected throughout the hydrological cycle. Although man's knowledge of the processes involved is now great, his ability to manipulate the hydrological cycle at other than a relatively local scale is limited by his ability to control global climate. He has achieved most success in regulating the amount of water passing through the soil. The two most far-reaching effects have been to cut down 'losses' due to interception and to evapo-transpiration, and to increase the proportion of surface run-off as a result of deforestation alone, of deforestation followed by grazing and burning, or of the replacement of a perennial natural vegetation cover by an annual crop which provides a limited cover for only part of the year. Losses are probably least in the 'small grains'—oats, barley and wheat—when row-drilling, combined with the vertical disposition of leaf blades, reduces interception and when the period of maximum foliage-cover may be as little as six weeks. On the other hand, as has been demonstrated in Kenya, replacement of a forest cover by a perennial tree crop with comparable canopy interception and rooting depth—such as tea and/or coffee—has been successfully effected without any significant change in the hydrological cycle. The knowledge that deforestation reduces significantly the consumption of water by plants and thereby increases what the water-engineer refers to as the 'water yield' (see figure 2.3) from the land, has led to efforts to manage vegetation of water-sheds,

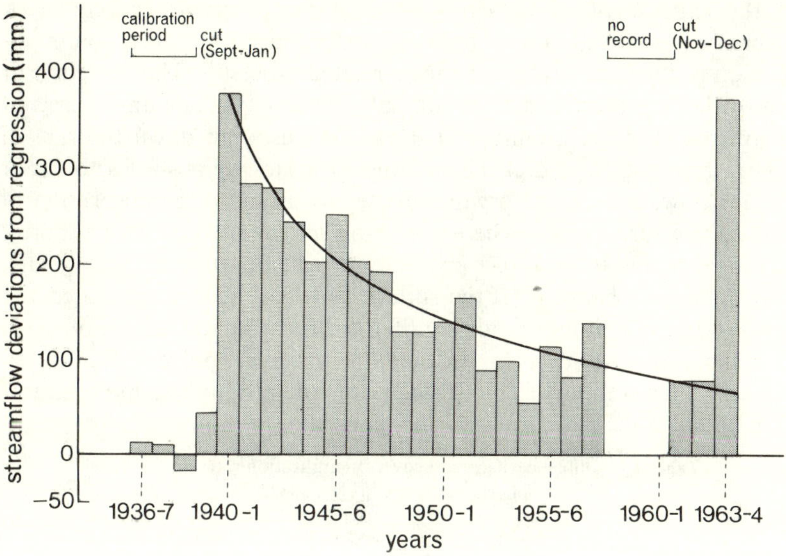

Figure 2.3 Increase of water yield after clear-felling a forest: a unique confirmation. After only three years of comparison of the monthly water yields of two small forested watersheds, one was clear-felled in 1940. Water yield increased by 373 mm (14·6 in) depth over the watershed. As the forest regrew, yields declined logarithmically with time. Twenty-three years later, the cut was repeated. The yield increase was almost identically repeated. (Diagram from Hibbert, 1967, and Pereira, 1973.)

or over the whole or part of a water-catchment area. The aim may be to increase the flow through surface and underground routes, which can then be tapped for uses other than agriculture, or, conversely, through the soil-plant route in order to mitigate the undesirable effects of surface run-off.

Deforestation can also, because of the reduction in the consumption of water, increase the amount of soil water. It may, dependent on permeability, increase the amount of water percolating through the soil and accelerate soil leaching or the washing out of the upper part of the profile of soluble minerals. Conversely, in less permeable material it may cause soil saturation and, in cool humid climates, lead to peat formation consequent on the resulting anaerobic conditions. The widespread development of peaty soils, for example, may well have been initiated by deforestation coincident with or after a change of climate from the drier and warmer conditions under which the forest originally developed, to the cooler, more-humid present.

Unfortunately, however, increase in *surface run-off* usually takes place at the expense of *infiltration* with even more serious consequences. Decrease in infiltration rates following deforestation or vegetation modification is due to several closely inter-related processes. The depletion or removal of a protective canopy not only increases the amount of precipitation reaching the ground, but it can seriously cut down the rate of infiltration (see Table 2.1). This is due to a lack of obstacles to check surface flow, and of soil organic matter to facilitate the absorption of surface moisture. Finally, where a bare soil surface is exposed for a shorter or longer period to the direct physical impact of particularly heavy rain, two things can happen: (1) the surface soil may become saturated too quickly and inhibit infiltration; or (2) the direct physical impact of heavy rain can cause '*soil capping*'. In the latter instance, pulverization of the surface material effectively blocks soil pores and produces an impermeable skin or cap which accelerates surface run-off.

Table 2.1 Influence of ground cover on infiltration rates on the same soil association (Chorley, 1969).

Ground Cover	*Infiltration rate (mm/hour)*
Old permanent pasture	57
Permanent pasture (moderately grazed)	19
Permanent pasture (heavily grazed)	13
Strip cropped land	13
Weeds or grain	9
Clean-tilled bare ground	7
Crusted ground	6

The amount of change in the relative balance of infiltration versus run-off can have serious and widely ramifying repercussions. It will obviously be one of the factors promoting accelerated soil erosion—a process which will be discussed in more detail later. It accentuates variations in river and lake levels, and spates (or rapid rises in water level) become more frequent as the time-lag between precipitation and run-off decreases. In areas which combine steep slopes and intense precipitation, the resulting flood hazards can affect an area out of all proportion to that originally modified or altered.

The impact on the hydrological cycle is all the more serious where climatic conditions in terms of the soil-water balance are near critical limits for agriculture; and more particularly in those parts of the world where low annual precipitation is accompanied by marked seasonality and variability, together with high evaporation rates. Increase in surface run-off at the expense of infiltration gives rise to a situation where the proportion of the precipitation retained in the soil drops, loss by direct evaporation from the surface increases, and natural or artificial withdrawal from ground-water reservoirs exceeds recharging. In time, as the general water-table falls, wells start to dry up, springs and streams cease to flow, and ground-water must be extracted by pumping from ever greater depths. Unirrigated crop yields fall. These symptoms of drought become more marked and more pernicious, while continued cropping and grazing increase the risk of wind erosion. Eventually the desert begins to encroach on formerly productive areas.

The increasing incidence of drought and the expansion of the desert margins are familiar phenomena both in northern and southern Africa. The decline in once-settled sites along the North African coast has been ascribed, at least in part, to such causes. In 1920 the South African Government set up a Drought Investigation Committee, and in 1951 a Desert Encroachment Committee. The latter was concerned particularly with wind erosion consequent upon the desiccation of soil stripped bare by widespread over-grazing in semi-arid areas. Desert encroachment, or what has been described as the 'saharization' of Africa, formerly thought to indicate a change of climate, is now generally accepted as a consequence of vegetation clearance and over-cropping in climatically marginal environments. Once initiated the process is both difficult and costly to reverse, particularly on the scale involved.

Micro-climatic modifications

Reduction or clearance of the natural vegetation cover has considerable effect on micro-climate. This is the condition of that part of the atmosphere directly above the surface of the ground. In general, the vegetation cover forms an insulating layer between the surface of the ground and the free atmosphere above. Its efficacy in this respect is a function of its *biomass* (or amount of living matter) which is dependent on the size and density of the constituent plants. The larger the plant biomass, the more drastic the change of micro-climate should it be cleared. The surface is exposed to the often brutal impact of high wind-force, direct insolation, and unimpeded precipitation. As has been noted before, evapo-transpiration losses are reduced. However, a combination of higher wind-speeds, lower atmospheric humidity, and higher temperatures consequent upon direct insolation, increases the rate of surface evaporation. In the absence of a 'vegetation canopy', the ground surface is exposed to full sunlight; its temperature, and that of the air above, rises more rapidly and to higher maxima during the day and in summer, but falls more rapidly and to lower levels at night and during the winter, because of unimpeded out-going radiation. The result is a replacement of a formerly sheltered equable micro-climate by a more violent and variable one in which the diurnal and annual ranges of temperature and humidity can be extreme.

The impact of vegetation clearance on micro-climate will obviously vary according to the nature of the pre-existing vegetation and to the prevailing climatic conditions. The effect of deforestation is usually greater than for lower forms of vegetation, and is nowhere quite so injurious as in the Tropical Rain Forest. In the absence of undergrowth characteristic of this dense forest, exposure of the ground-surface to constantly high temperature and direct insolation results in a very rapid increase in the rate of decay of any unhumified organic material. The effect of less dense woodlands and grasslands on the micro-climatic conditions is not so great as in the Tropical Rain Forest. Nevertheless, exposure of the ground to the unmitigated effects of alternating wet and dry seasons can result in serious soil desiccation in savanna woodland and grassland. At high latitudes, where permafrost or permanently frozen ground (often the legacy of a previously colder climatic regime) underlies the forest cover, removal of the latter can have the effect of lowering the level at which the permafrost occurs. It has been noted that the clearance of trees and shrubs in Alaska (near Fairbanks) resulted in a lowering by 1·8 metres in five years and 2·4 m in ten years; when all the vegetation was stripped, the level of the permafrost dropped 2·7 m in five years, 3·8 m in ten years. This was a consequence of higher soil temperatures following direct surface insolation during summer. Conversely, however, the maximum depths of seasonal freezing were increased in the winter. The resulting change in balance between permanently and seasonally frozen ground has direct repercussions on soil and ground stability. The lowering of the permafrost level, in particular, incurred a gradual and permanent ground subsidence.

Increasing knowledge of the process involved has enabled man deliberately to create or modify micro-climatic conditions for agricultural purposes. The use of trees to provide wind-breaks or shelter-belts in extremely exposed areas, or to protect particularly delicate crops, is a case in point. Decrease in wind speed cuts down the risk of physical damage to crops. In addition, it is accompanied by reduced evapo-transpiration and crop yields significantly greater within than beyond the zone of maximum shelter; excessive losses in animal body weight can be avoided.

The shelter-belt plays an important role in such diverse agricultural systems as upland stock rearing in Britain, horticulture and fruit growing, and extensive wheat farming on the vast exposed prairies and steppes of the USA and the USSR respectively.

Trees are also used to provide the shade necessary for the optimum production of crops such as tea and coffee. Some species may be planted merely as 'nurse-crops' to provide sufficient shade to suppress vigorous undergrowth that might check the development of other seedlings. In Britain the lodge-pole pine (*Pinus contorta*) has an initial rapid growth that makes it particularly suitable for this purpose. In the case of valuable and frost-sensitive citrus fruits, often quite sophisticated measures are taken to protect the crop from risk. These include the use of 'smudge pots' in which smoke-producing material is burned in order to cut down heat losses from the ground surface by out-going radiation on clear still nights; and the installation of electrically operated 'wind-fans' to ensure constant air circulation and thus prevent cold air draining down surrounding slopes from settling on the ground. Only in the glass-house, however, can micro-climatic control be complete.

Soil erosion

Cultivation

In most cases clearance of a pre-existing vegetation cover has been a preliminary to cultivation or soil management for crop production. Cultivation involves a cycle of processes which include the breaking-up and turning-over of the surface soil to a depth which may vary from 40 cm to a few centimetres, by ploughing, discing, harrowing and rolling, followed by fertilizing, seed-drilling, weeding and eventually harvesting. The basic aim of all methods of physical cultivation is to produce a seedbed with as good a *tilth* (i.e. physical condition or *structure*) as possible. The term *tilth* or *structure* is used to describe the way in which the individual mineral particles are arranged in the soil—whether they are completely discrete and loose as in sand, or are more or less firmly bound together in compound particles or aggregates. The latter can range from very small crumbs or granules to large 'clods'.

The most favourable condition for plant growth is that with a well-developed stable 'crumb' structure in the upper rooting zone of the soil. This is because it provides the optimum balance of (*a*) large pore spaces between the 'crumbs' which facilitate free drainage and good aeration, and (*b*) small pore spaces within the crumbs helping to retain water and nutrients for the growing plant. The type and stability of structure in any cultivated soil is dependent partly on its original constitution (the size of the mineral particles and the amount of organic matter, which with the finer clay material are important binding agents, the movement of water and minerals through the soil, the type and number of soil animals, etc.) and partly on the way the soil has been managed. In some cases cultivation builds up and maintains a good structure, in others it breaks down and destroys structural stability with, as will be seen, unfortunate consequences.

In all types of cultivation, however, the soil is disturbed and laid bare for a longer or shorter period. This inevitably introduces the risk of *soil erosion* (or *accelerated soil erosion* to distinguish it from 'natural' or 'normal' erosion). Erosion or the removal of weathered mineral debris by such agents as gravity, wind, water, or ice is a natural process. It is characteristic of areas where the surface is not completely protected by vegetation, as in deserts, on steep slopes, along river courses and coasts. On relatively level terrain, with complete vegetation cover and well-developed soil, erosion may be so slow as to be imperceptible. It is then assumed, theoretically, that a state of equilibrium has been attained, any losses from the surface being made good by the gradual addition of freshly weathered mineral material at the base of the soil. If the equilibrium is disrupted by a quantitative or qualitative change in one of the variables concerned, removal of the surface soil will take place at a rate faster than it can be replaced. This is soil erosion which, once initiated, can within the space of a few years destroy a soil which has taken centuries to form.

It is now widely accepted that all the visible effects of accelerated soil erosion are a result of man's activities. Indeed, it has been described as one of the most vicious and destructive forces ever released by man. Further, activities other than agriculture which cause soil erosion are of little significance. According to Butzer (1974), 'it has posed a latent if not chronic environmental problem ever since agriculture became the dominant mode of subsistence in the Old World 10 millenia ago'. This author also suggests that, before the sixteenth century A.D., soil erosion was limited in extent and confined to Eurasia. Primitive cultivation with the hoe or digging stick was concentrated on small plots on level land, and hence would have had a limited impact on the soil. Despite the introduction of the ox-drawn scratch-plough (*ard*) in the third century B.C., there is little evidence of significantly widespread erosion in Temperate and Mediterranean Europe until about 1000 B.C. For a long period, however, soil erosion was probably confined to the most heavily populated areas of the Mediterranean Basin where it is still an endemic and chronic problem. Since the sixteenth century, population growth combined with rapid technological developments have contributed to more widespread and generally ruthless soil exploitation, particularly in formerly uncultivated areas of the world, and soil erosion has become rampant.

Man's perception and understanding of the processes have been slow to develop. Although soil erosion and its possible causes were noted by early classical writers, such as Plato, systematic investigations were not undertaken until the end of the nineteenth century in Germany. An awareness of the real nature and global extent of the process did not emerge until the 1930s when drought and the dust-bowl conditions of the south-west United States were popular topics of novels and textbooks, and *soil erosion* was as emotive a term as *pollution* today. The processes and control of soil erosion became a major area of research in the 1940s.

Causes of soil erosion

The two principal causes of soil erosion are: (1) the degeneration and/or removal of a formerly continuous vegetation cover by deforestation, overgrazing and/or burning, and cultivation; (2) a reduction in soil stability due to a progressive decrease in organic-matter content and a consequent loss of structure. The first exposes soils, the second makes them susceptible in varying degrees to accelerated erosion. The amount and type of soil erosion are functions of what have been called the *erosivity* of the agents involved and the *erodibility* of the soil.

Agents of soil erosion

The two main agents of soil erosion—water and wind—are not mutually exclusive, though they tend to dominate different climatic regions (see figures 2.4 and 2.5).

Water erosion

Water erosion is more serious since it is most prevalent in areas actually or potentially cultivable. Its effect is dependent on the intensity or rate of precipitation, and is primarily a function of rain-drop size. In the absence of a protective vegetation cover, the velocity of falling rain is unchecked by interception, and its impact on the ground is therefore greater than would otherwise be the case. Also, as rainfall intensity increases, so infiltration rates decrease. It has been calculated that rain falling at the rate of 2 tons per hectare will be absorbed by a given soil at the rate of 20 cm per hour; 66 tons per hectare on the same soil at only 2 cm per hour. As a result, the amount of surface run-off increases; and on sloping ground the water velocity and hence its erosive power are further enhanced.

The significance of rain-drop impact in soil erosion was not fully realized until the late 1940s, when Ellison demonstrated the nature of *splash erosion.* The kinetic energy released on impact is sufficient to break down

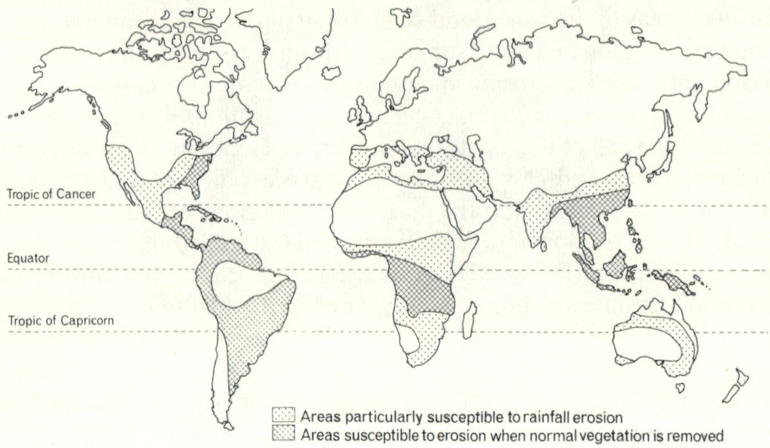

Figure 2.4 General distribution of rainfall erosion.

or pulverize the soil aggregates, separate individual soil particles, and project them into the air. It has been calculated that 5 cm of rain falling at 36 km per hour could raise 15 cm of top-soil 1 metre into the air if it all fell at once! This 'kick' is obviously proportional to the size and velocity of the rain drop, and Ellison noted that 'the impact of rain-drops in a violent storm may blast more than 200 tons/ha into the air'. However, the lateral movement of soil particles can be four times as great as the

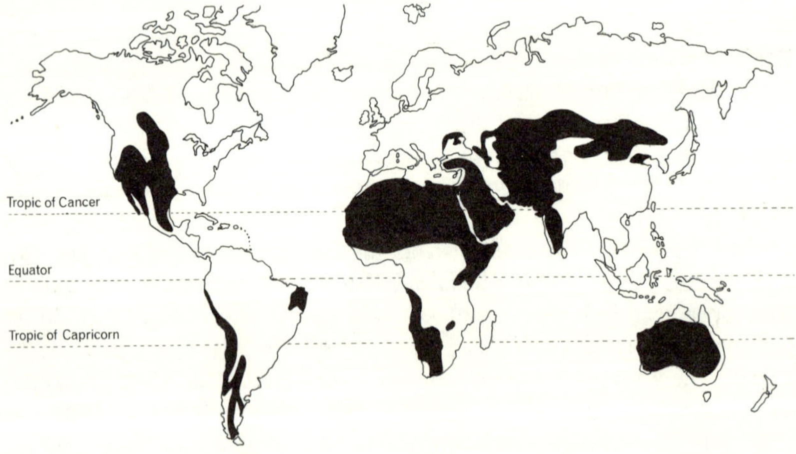

Figure 2.5 Arid and semi-arid regions of the world susceptible to wind erosion.

vertical movement and, on sloping ground, splash will be accompanied by a predominantly downhill movement. This particular process, which can account for up to 90% of all soil-erosion loss, tends to remove the top soil evenly over an affected area as what is called *sheet-wash*. In doing so it gradually smooths and levels ground by removing soil from the top and depositing it at the base of a slope.

Splash erosion, then, triggers off a chain of damaging reactions which include: (1) *structural deterioration*, with finer particles capable of either surface removal or of leaching down the soil profile; (2) *surface-sealing* of pores by pulverized soil particles which, as has already been noted, aggravates surface run-off; (3) *piling* and *burying* top-soil under eroded material; (4) *elutriation* or *nutrient depletion* as a result of the increasing ease of transport of finer particles of clay, silt and organic matter. This process can be particularly invidious on the sandier types of soils, because the effect of the removal of fine material and the consequent deterioration in the water-holding capacity of the soil will be proportionately greater than in the heavier loamy and clay soils. In some badly eroded areas, agricultural soils have been elutriated to the extent that they now resemble beach sands.

With increasing rainfall amounts, sheet-wash tends to be replaced by *scour erosion*. Run-off becomes concentrated by initial surface irregularities into a distinct flow which cuts temporary rills or more permanent gullies. The former are usually not deep enough to impede, or persist after, cultivation. The latter are more localized, particularly on long slopes where channelled run-off becomes increasingly concentrated at the base of a long slope or in a depression. Although gulleys and ravines are the most obvious and spectacular types of erosion damage, their contribution to total soil losses is relatively insignificant, in terms of volume of material lost or fertility of areas affected, compared to the other more insidious though less easily detected forms of water erosion.

Wind erosion

When exposed soils dry out, they are particularly susceptible to *wind erosion*. The primary process in this case is *saltation*. This occurs when the wind force is sufficiently strong to project fine soil particles into the air by literally bombarding them with coarser particles. It usually involves particles, either discrete or aggregated, in the size range of fine sand (0·2 mm) to coarse silt (0·1 mm). Those less than 0·1 mm can, once projected into the air, be carried by suspension, whereas those over 0·5 mm are moved only by traction or rolling along the surface. Fine sandy soils, calcareous clays and peat soils, with over 60% unaggregated particles in the size range 0·1–0·05 mm are most liable, when dry and unprotected, to blow; those with less than 40% are much less liable to damage. The result of soil-blowing is the gradual and selective removal or *deflation* of finer particles. The agronomic consequences are very similar to those associated with elutriation by water. Although soil erosion by wind is most widespread in areas where low annual rainfall is combined with a high degree of variability and a marked seasonal incidence, it can also occur as a result of certain localized conditions in humid areas.

Imbalance of soil condition, land use and management

While the type of soil erosion is generally a function of climatic conditions, erodibility is very dependent on soil condition, slope and, even more importantly, on land use and management. Indeed Hudson (1973) maintains that the difference in intensity of erosion resulting from variations in the management of the same type of soil can be much greater than from variations in the erodibility of different soils under the same type of management. Within any particular land-use system there can be wide variations in degree of soil erosion resulting from cultivation practices associated with individual crops (see Table 2.2).

Table 2.2 Measured rates of soil erosion from experimental plots (Chorley, 1969)

Area	*Mean annual rainfall (mm)*	*Vegetation cover*	*Per cent run-off*	*Soil loss (mm/year)*
South-east USA	2500–4000	Oak forest	0·8	0·008
		Bermuda—Grass pasture	3·8	0·030
		Scrub oak woodland	7·9	0·10
		Barren abandoned land	48·7	24·4
		Cultivation-rows around contour	47·0	10·6
		Cultivation-rows downslope	58·2	29·8
Rhodesia	1000	Dense grass	2·7	0·018
		Bare ground	38·0	2·3

Land-use systems which accelerate erosion are those which involve an extractive or 'robber' type of soil exploitation: More nutrients are extracted from, than are returned to, the soil. This drain is usually accompanied by a gradual reduction in amount of organic matter, deterioration of soil structure, modification of the soil water-balance, and instability which is the precursor of erosion. This process can be 'direct' in the case of arable cultivation, or indirect in that of livestock farming based on the uncontrolled use of open-range or rough grazing provided by the existing vegetation cover. Grazing by domestic animals tends to be selective. Also more nutrients are lost in the crop than are returned by animal faeces and, under conditions of over-grazing, the available forage may be eaten down at a rate greater than it can be replaced by regrowth. Periodic firing of the vegetation, the traditional method of management, helps to speed up nutrient cycling and promote new shoots; but in many cases nutrient loss (as a result of the vaporization of nitrogen or the leaching and erosion of ash) probably exceeds the return. This leads to a progressive decline in soil fertility and hence in the amount, productivity and nutrient value of the herbage. Long-continued grazing and burning inevitably weaken the vegetation cover, and eventually lead to the exposure of the underlying mineral soil with often disastrous results, particularly in areas with heavy seasonal rain or on mountain slopes everywhere. Soil erosion promoted by grazing is not only long established but it occurs on a scale that far exceeds the direct impact of arable cultivation. In view of the limited alternatives for the use of the traditional grazing areas of the world, possibilities of mitigating its effects are as yet slight.

Before the mid-eighteenth century, most cropping systems tended to be extractive, except in the case of intensively cultivated and manured garden-types of agriculture or paddy rice cultivation. Resting the soil or leaving it fallow for a period of time was the traditional method of restoring fertility common to a great variety of systems. Soil erosion consequent upon long-continued cultivation was, as far as is known, almost totally restricted to steep slopes of uplands adjacent to the highly populated lowlands around the Mediterranean. The opening-up of the formerly uncultivated areas in temperate North and South America, Eurasia and Australasia, provided apparently unlimited land resources at a time when agricultural machinery was becoming more efficient and world demand for grain and industrial crops was growing rapidly. Commercial, and increasingly specialized, agriculture was characterized by an extractive and wasteful exploitation of virgin soils; this initiated soil erosion at a rate and on a scale hitherto unknown. Mixed farming, based on the systematic rotation of crops and grass common in Europe, was abandoned for specialized monocultures—the cultivation of nutrient-demanding cash crops such as tobacco, cotton, sugar cane, or maize, on the same land for several successive years, usually with little or no fertilization. Declining fertility eventually bore witness to the depletion of organic matter, and the structural deterioration of the soil. Those initially poor soils succumbed all the more easily to the intense rainfall of continental summers in the United States, and even more rapidly where slopes accelerated run-off. However, the declining fertility which was early associated with particular crops is now thought to have been the result rather than the cause of soil erosion. The latter is a function not so much of the nutrient demands these crops make on the soil, as of the methods of cultivation associated with certain types of crops.

The 'erosion-promoting' crops in most cultural systems are usually annuals grown in regular, often widely spaced, cleanly weeded rows. A relatively short growing-season, combined with an incomplete cover even at maximum growth, and a long fallow period means that some bare disturbed soil, lacking organic matter, is exposed virtually throughout the year. Row cultivation aligned downslope at gradients of over 8° in temperate, or as little as 2° in tropical, areas invites disastrous erosion.

Erosion and monoculture

Some of the most extensive and spectacular soil erosion has been that which resulted from the extensive monoculture of wheat in semi-arid regions of the United States and Canada in the 1930s and, more recently, in the USSR in the 1960s. In periods of high demand, cultivation was pushed into areas where the rainfall is marginal for arable farming and the risk of often extended periods of drought is high. In this event, exposed and disturbed soils are particularly susceptible to wind erosion. Such were the conditions which, in the drought-stricken 'thirties, caused widespread devastation over the area from Kansas across south-western Oklahoma to western Texas, which is still remembered as 'The Dust Bowl'; and it was this catastrophe which shocked the whole world into a full realization of the nature and extent of soil erosion.

More recently, the introduction and expansion of cereal monoculture has, in certain parts of Britain, led to soil-blowing on a scale not usually associated with a humid climate. An inquiry by the Agricultural Advisory Council in 1973 revealed that as a result of over-cultivation and monoculture, combined with the exclusive use of artificial fertilizers and a deficiency of organic matter, light sandy soils tend to dry out excessively during drought periods. Incapable of forming and maintaining stable aggregates, the risk of wind erosion has been greatly increased, and localized dust storms have become more and more frequent in the arable areas of the Midlands, eastern England and parts of Warwickshire. The effectiveness of wind erosion has been further exacerbated by the removal of hedges, and other such potential wind-breaks, in order to increase field size in the interest of more efficient use of machinery.

Erosion of tropical soils

Humid and sub-humid tropical soils are more liable to erosion than temperate ones, making agriculture peculiarly precarious. In the first place, despite the luxuriance and rapid growth of tropical forest and savanna vegetation, the amount of litter accumulating on the surface (and of organic matter in the soil) tends to be relatively small and limited in depth. This is a consequence of rapid decomposition. Destruction of the protective vegetation cover speeds up the process, and ensures a rapid depletion of the initially small reserves in the soil. In addition, most tropical soils, particularly those associated with acidic parent-material and gently-sloping terrain, have been subjected to a much longer period of weathering and leaching than in temperate regions. In the latter, soil formation was interrupted either directly or indirectly during the Pleistocene glaciations. Hence many tropical soils are naturally nutrient-deficient, but rich in residual iron and aluminium compounds. While they are often porous and friable, their capacity to retain water and nutrients is variable and low.

Until relatively recently, however, the risk of soil erosion as a consequence of the intensity of tropical rainfall has been mitigated either by the natural vegetation cover which still persists over considerable areas or by traditional methods of cultivation which minimize soil exposure and disturbance (figure 2.6). One of the commonest land use systems is *shifting agriculture*, very similar to that practised in Britain in Neolithic times, some 5000 years ago. Cultivation is based on small areas of cleared forest-land from which all but the larger tree stumps are removed by cutting, ring-barking and burning. Soil preparation is limited by the primitive tools available, such as the hoe and digging-stick, which do little more than scratch the surface. However, they have the great advantage of preserving soil moisture and organic matter. In addition, the common practice of planting an intimate mixture of seed and tuber plants on mounds ensures a year-long protective crop-cover. Cropping of any one patch is continued for only about three years, after which it is abandoned in favour of another clearing. Provided disturbance has not been too great, the recolonization of the old clearing will be rapid, the scar will heal, and a process of natural regeneration restores the depleted soil fertility. Shifting agriculture is, therefore, based on a long rotation of tropical mound-cultivation and a long tree-fallow period.

Unfortunately two events have within a relatively short period upset the delicate balance achieved between this low-yield but essentially fertility-conservative type of farming and the physical environment, and have caused a rapid acceleration of soil erosion. One has been increasing population pressures on dwindling land resources, with a concomitant lengthening of the cropping and a shortening of the fallow period. This has resulted in a gradual degradation of the vegetation from forest to woodland, and eventually to grassland, with increasing exposure to high-intensity rainfall. The second event has been the application of temperate agricultural techniques to the essentially fragile tropical soils. None has been more catastrophic than the impact of the tractor-drawn steel plough. Exposure and disturbance of tropical soils will cause a greater change in a shorter time than in temperate soils. In the tropics, considerable areas,

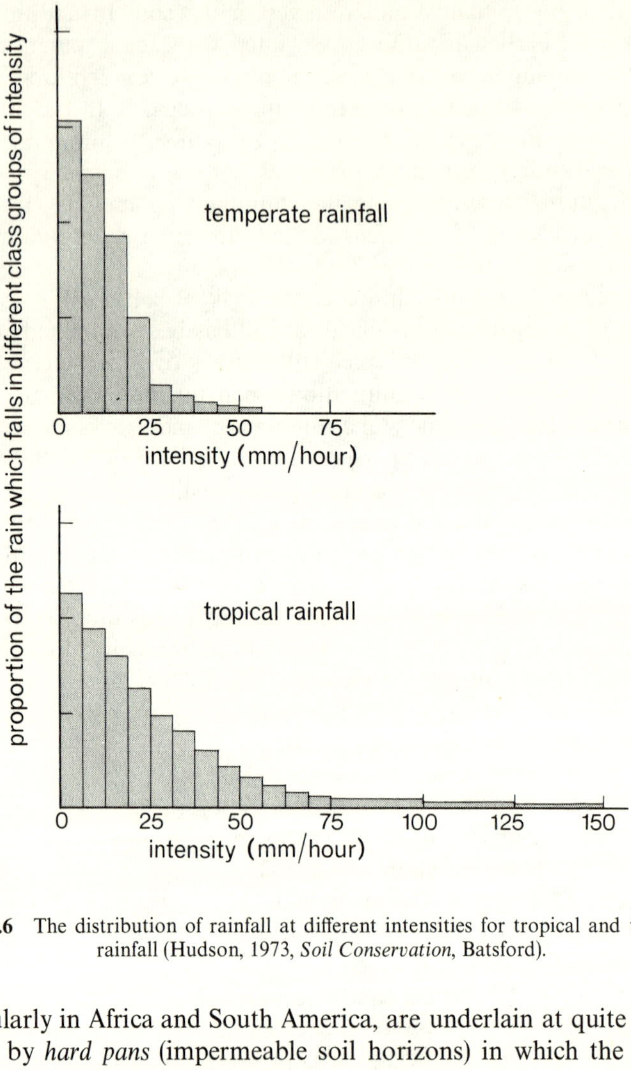

Figure 2.6 The distribution of rainfall at different intensities for tropical and temperate rainfall (Hudson, 1973, *Soil Conservation*, Batsford).

particularly in Africa and South America, are underlain at quite shallow depths by *hard pans* (impermeable soil horizons) in which the mineral matter has become cemented together, usually by iron-oxide. Removal of the overlying soil exposes these pans at the surface, where they form hard surface crusts. In other cases, the soil may be so oxide-enriched as to be full of concretions; nevertheless, it will be porous and friable under natural conditions. However, when exposed by clearance and cultivation, it becomes hard and indurated as a result of desiccation. The end result may be bare sterile rocky land-surfaces to which the French have given the name *bowal*.

Consequences of soil erosion

The immediate consequences of soil erosion then are nutrient losses, particularly nitrogen, phosphorus and sulphur, which tend to be concentrated in the upper part of the soil profile. This 'fertility-erosion', as it has been called, can cause loss of plant nutrients comparable in magnitude with the removal of the same elements by a harvested crop. Further, soil desiccation and the exposure of little-weathered parent material or completely sterile pans can occur. Severe gulley erosion can seriously impair the workability of land. In addition, the effects of soil erosion can extend far beyond the area directly affected. As a result of the addition of mineral and organic sediments to surface water run-off, it was probably the first and most widespread source of water pollution. Addition of sediment to water increases its turbidity and the increased load is eventually re-deposited elsewhere. Silting can damage fish-spawning beds and seriously block water courses and reservoirs. Sediment-polluted water can, as a result of abrasion, increase the wear and tear of machinery. It can disrupt irrigation schemes by interfering with infiltration rates and increasing evaporation losses. Indeed, as Pereira (1973) has shown, relatively small areas of extreme erosion-damage can have repercussions over a disproportionately large area. He describes the case of the Parana River watershed in the semi-arid High Andes of South America. Here soil erosion only affects 4% of the area. Studies have demonstrated that some 80% of the 100 million tons of sediment carried downstream comes from the over-grazed area at the head of the Bermejo tributary (where the Argentinian cattle industry was founded some 400 years ago). This ever-increasing sediment load has had a serious economic impact far downstream as a result of silting around the port of Buenos Aires. That this severe soil erosion is an accelerating process is indicated by the growth of the delta at the Parana mouth: in 1873–97 it was advancing at a rate of 46 mm/year; in 1900–60 at 84 mm/year with no increase in river flow.

Rehabilitation and prevention

During the past fifty years or so, a number of agronomic techniques have been developed, either to rehabilitate damaged areas, or to reduce and combat future erosion risk. Some aim to cut down soil losses by ensuring adequate surface cover, by rotating or inter-planting crops with high cover-value plants such as grass and certain forage-crops, or by higher plant densities as a result of closer drilling or increased fertilization. Stubble-mulching and trash-farming were developed in the United States in response to the need to protect the surface of arable fallows, particularly in dry-land wheat farming which is based on a rotation of a one-year wheat with one to three years fallow. In the case of the former crop, residues are chopped up and spread on the surface. In the latter the crop residues are ploughed into the soil immediately after spreading, and cultivation then follows in the normal manner.

Other methods of erosion control include *minimum tillage* or even *no-till* systems. Minimum tillage aims to reduce the number of operations, and hence the amount of soil exposure and disturbance, by planting directly after ploughing. In recent years, however, many farmers in the south-eastern United States have returned to what is in fact a primitive method —that of cropping without ploughing. Cash crops are drilled directly into cover crops, such as winter rye or grass, which are then killed by Paraquat. The land is never left without the protection of a crop cover and its soil-retaining roots. As the old crop dies down, it in turn forms a close mulch around the next crop, keeps down weeds and helps to retain soil moisture. No-till systems eliminate run-off and the instability of soil structure produced by the growth of crops, since top soil is never disturbed and conventional inter-row cultivation of the crop is also eliminated. Other methods, designed particularly to check excessive run-off on sloping ground, have been to replace down-slope ploughing with contour ploughing, so that the plough furrows lie parallel, rather than at an angle, to the maximum slope of the ground. Contour ploughing is frequently combined with strip-planting, i.e. planting different crops in alternating rows, so as to reduce the amount of ground left bare at any one time.

Most important, however, are those techniques aimed at structural rehabilitation of the soil, and particularly the build-up and maintenance of a stable crumb condition already mentioned. They include the addition of organic matter (or humus)—one of the most effective binding materials —together with lime, to the soil. The latter encourages flocculation of the fine clay-sized particles. More significantly, however, it counteracts acidity and maintains the soil at a pH which is most favourable for organic decomposition *and* for soil animals. Of the latter, the most important in temperate areas are earthworms. It is now generally accepted that worm-worked soils contain considerably more water-stable crumbs than those without worms, though the reasons why this should be are not yet fully understood. Of equal importance is the presence of the roots of grasses and clovers, since their densely ramified root-mats add a large amount of nitrogen-rich humus directly into the soil and promote crumb formation. In some exceptionally favourable circumstances crumb structure can be renewed within four to five years under grass in Britain, though it can require up to fifteen years.

The need to combat soil erosion in the New World saw the introduction of cover-crops which would protect the surface and build up the depleted organic matter reserves in the soil. These included the soya bean and kudzu (a type of Japanese vine) both introduced into the United States in the 1930s specifically for erosion control—as well as alfalfa (lucerne) and sown grasses. They gradually began to replace the fallow break in the monoculture of wheat, maize and cotton. They became the basis of crop rotations designed (as they originally had been in eighteenth-century Britain) to conserve soil fertility and stability; they initiated new and more diverse cropping systems, or livestock and crop combinations in formerly one-crop areas. In Britain, the trend towards arable monoculture in the 1950s was stimulated by high grain prices and the use of artificial fertilizers and herbicides. Traditional rotations began to be replaced by five to six years of barley, interrupted by only one or two years of another cereal, usually wheat. Already, however, deterioration in soil structure, consequent upon the use of heavy machinery and a depletion of soil organic matter, combined with the spiralling costs of artificial fertilizers, are encouraging a swing back to the traditional and more fertility-conservative mixed rotation farming.

In the tropics

In contrast to temperate climatic regions, erosion control and rehabilitation have made relatively little progress in the tropics. This is partly because of a lack of capital and technical expertise. More significantly, however, it is because methods of erosion control, developed within the framework of temperate commercial farming, are no more applicable than are the associated methods of cultivation to tropical conditions. The rate of erosion on the naturally infertile and fragile soils is more rapid than in temperate regions. Further, erosion damage is more likely to be virtually irreversible, at least under existing technical and economic conditions. The risks involved in ploughing tropical soils cannot be significantly reduced by modifications of cropping and cultivation, such as have proved so successful in the United States. Neither the mineral constitution of the soils, the soil animals nor the types of grasses native to these areas are conducive to the development of a crumb structure comparable to that formed under temperate grasslands. Indeed, as has already been pointed out by many writers, the traditional minimum-cultivation techniques of the primitive systems of shifting agriculture appear to be those best adapted to tropical conditions, and to be more successful in controlling soil erosion. The French geographer Gourou (1974) maintains (and not without reason) that, except on inherently fertile alluvial and volcanic soils, the only ecologically successful form of modern commercial farming over much of the humid tropics of Africa and South America can be that which replicates as closely as possible the natural vegetation and soil conditions. This would obviously favour the tree rather than the field crop, plantation agriculture rather than arable or stock farming.

The net global result is that areas where soil erosion has been reduced are insignificant compared to those in which it has increased either in rate or extent.

Soil water balance

In addition to seed-bed preparation, regulation of the amount of water in the soil has long been one of the most important aims of cultivation. Over a long time man developed techniques of coping with a surplus or a deficit of soil water, either of which conditions will reduce crop yields. *Irrigation* has been accompanied by quite fundamental modifications of the physical environment over considerable areas. The origins of irrigation are ancient, and irrigation agriculture formed the basis of early stable highly-developed civilizations in Mesopotamia, Persia, India, China and Peru. Today irrigated land accounts for about 13% of the world's arable area with the particularly high proportion in Asia (see Table 2.3) being accounted for by the more primitive uncontrolled (i.e. rain-fed) irrigation of the rice paddy fields. Technical developments which have facilitated the extension of the more modern schemes have been mainly concerned with improvements in methods of water storage, diversion and application.

Table 2.3 World distribution of irrigated land (Cantor, 1967)

	Percentage of total irrigated land	*Area of irrigated land (millions of acres)*
Asia	64·8	239·8
USA	10·0	37·0
USSR	8·3	30·7
Europe	5·9	21·8
Africa	3·8	14·1
Central America	3·4	12·6
South America	3·2	11·8
Australasia	0·6	2·2

Irrigation involves the diversion of water from an area of supply to one of need. The latter may be due to very low and uncertain rainfall in arid or semi-arid regions, or to a marked dry season in areas where temperatures are otherwise high enough for crop growth. Irrigation is also used to make up slight water deficits during the growing season—as in the summer months in south-east England—in which case it is *supplemental* to an existing rainfall supply. Irrigation thus effects a major change in the hydrological cycle of relatively vast areas of land. Water which might otherwise have been evaporated from open water, or have found its way back to the sea by either surface or ground-water flow (see figure 2.2) is diverted through the soil-plant route. Irrigation, therefore, involves a heavy use of water, and usually constitutes a loss to other potential non-agricultural uses.

Problems resulting from irrigation

Both the diversion and the application of water have created a number of environmental problems which have not yet been completely solved. The two sources from which water is diverted are *ground water* and *surface water*. Irrigation started with the diversion of water from rivers and lakes. Surface water is still the major source, though methods have been developed to store as much surface water as possible against future use in reservoirs of varying size and complexity, to divert it over greater and greater distances by canal and aqueduct, and to supplement surface water in one river catchment by major pipeline diversions from another. Such storage and transfer of water has been accompanied by evaporation losses which are not only considerable, but are also greater than would otherwise have been the case. Extraction from surface water has also resulted in reduction in river flows downstream, or in river basins from which water has been diverted. Where river flow is from more humid to more arid conditions, as in the case of the Colorado or the Indus, reduction in volume in areas of high potential evaporation can cause a marked increase in the salinity of the water, and an increasing deposition of silt on the river bed.

Extraction from ground water by free-flowing wells, or by pumping, developed later, but the physical impact is greater than in the case of surface-water diversion. Modern irrigation schemes based on ground-water are usually in desert areas, e.g. south-west California and north-west India, and often put a very heavy demand on limited resources. Extraction inevitably exceeds the rate of ground-water replenishment, either by rain or by underground seepage. The result is usually a gradual fall, followed by an increasingly rapid fall, in the water-table and a depletion of stream flows. Unrestricted pumping of ground water in the Salt River Valley in Arizona lowered the water-table there by 70–90 cm in ten years. Over-pumping and the lowering of the general water-table has caused serious damage. Withdrawal of considerable volumes of water has led to land subsidence. This is now widespread in the San Joaquin Valley, California, where the rate of subsidence in the decade 1950–60 was as much as 24 cm per annum. Land subsidence undermines buildings and transport systems, as well as disrupting drainage and irrigation, and any other underground pipe or cable-lines. It can also bring about a permanent reduction in the capacity of the subterranean water-holding strata (i.e. the *aquifers*) themselves, as a result of direct compression and compaction by the weight of overlying rocks. In coastal areas, if the water-table level drops below that of the sea, the incursion of brackish or saline water can render ground supplies unsuitable for irrigation; and some of the most valuable products of modern commercial irrigation—the citrus fruits—have a very low tolerance of salt. Such problems are most severe in areas or countries with a heavy dependence on ground water for irrigation, e.g. Tunisia (95%), Morocco (over 75%), Israel (70%) and Saudi Arabia, where it is the only source. While extraction and diversion of water have had a direct and often harmful effect on the hydrological cycle in arid and semi-arid areas, the application of irrigation water primarily affects the soil condition.

The most efficient means of water application is by spray and, in modern irrigation schemes, great attention is given to producing that size of water-droplet that will reduce the physical impact on the soil, minimize evaporation losses and maximize penetration. Many schemes, however, apply flowing water direct to the surface. As it spreads out, this water sorts the surface soil particles. The fine clay and silt particles remain longer in suspension than the coarser fractions and, when eventually deposited on the surface, they clog up pores, cut down water infiltration rates and increase evaporation loss. Application of surface water, and particularly over-watering, can quickly raise the water-table and in extreme cases water-logged soils can result. In the Twin Falls area of the Snake River plain (Idaho) it has been calculated that, during the first 25 years of irrigation over a 100,000-ha tract of land, the water-table rose by as much as 100 m.

A high water-table impedes the downward drainage and leaching of soluble salts. Hence, in arid climates, a rise in the level of the water-table combined with high evaporation rates can cause *soil salinization.* This is the cumulative precipitation of salts such as the chlorides, sulphates, carbonates of sodium, calcium and magnesium near to or on the soil surface. It is revealed by whitish deposits to which the term *white-alkali soils* is applied. Crops vary in their tolerance of saline conditions. Over certain levels, yields will decrease to an extent that makes further cultivation uneconomic, if not biologically impossible. It is often necessary to pump off excess water in order to lower the water-table, to facilitate free downward drainage through the soil, and the flushing-out of excess salts. Unfortunately in soils rich in sodium this very process can cause an increase in soluble sodium carbonate in the soil water, which produces conditions so alkaline (pH9 or over) as to be practically biologically sterile. *Soil alkalinization* is extremely difficult and expensive to cure. It is somewhat ironic that all irrigation systems, because they are so very delicately balanced, contain within themselves the seeds of their own destruction. Indeed, it has been estimated that the amount of land ruined by irrigation now equals, if it does not in fact exceed, that potentially irrigable.

Drainage

Efficient drainage is essential to all successful agricultural operations, and an excess of soil-water is a major factor limiting cultivation and crop yields in many parts of the world. It is particularly characteristic of cool humid regions, such as western Europe, where the amount of rainfall exceeds the potential evapo-transpiration rate for all or most of the year. Under these circumstances it is further exacerbated by (*a*) lack of slope and low rates of surface run-off, (*b*) fine-textured silty and clayey water-retentive soils, (*c*) shallow soils lying on impermeable bedrock and (*d*) a high water-table. In low-lying coastal areas, as in the Netherlands, and flood plains, such as those of the Po Valley and the even greater Mississippi River, a high water-table alone can cause surplus soil water over very extensive land areas. Among the most intractable drainage problems are those of peat and peaty soils. These have formed where soil saturation has resulted in an oxygen deficiency which precludes the presence of aerobic bacteria. The rate of decomposition is retarded, and partially decomposed organic matter accumulates. Where peat forms in ill-drained basins or hollows, it can reach 20 m or more in depth. Peat is capable of absorbing and holding an amount of water up to three times its own volume, and drainage of such deposits tends to be accompanied by more drastic physical changes than in the case of a mineral soil. When peat is drained it shrinks and, as a result, the ground surface subsides. It has been estimated that drainage of peat land in Florida has caused subsidence at the rate of approximately 1 m in 24–36 years; and peat lands in the Central Valley of California have subsided by as much as 5 m since they were first drained for cultivation. Nearer home, shrinkage has similarly affected the nutrient-rich peat soils of the East Anglian Fens. In upland areas in Britain, deep-ploughing and drainage of shallower peat or peaty soils has been extensively carried out, primarily in the interests of either grazing improvement or afforestation. In some cases this may have accelerated run-off and increased spates in the rivers, though in the absence of reliable long-term data this remains a debatable point. In others drainage and concomitant drying has led to intense localized peat erosion, particularly in the uplands of north-east England and Scotland.

Impact of machinery on drainage

More important probably than the effect of drainage is the impact of modern methods of cultivation on the drainage capacity of the soil. These have involved the use of ever-larger heavier and more expensive machinery, in order to off-set declining man-power. The use of machines, however, has not been without physical problems, notably in all-arable farming and minimal-cultivation systems—particularly those denied the rest to the soil afforded by the grass clover break or its equivalent.

Machinery affects soil by compression, shear, slip, bounce and vibration. The effect tends to be least when soils are in a well-drained condition and with a soft or harsh consistency. Poorly drained soils, particularly the heavier clays and silts, can, however, quickly become plastic if subjected to undue pressures. Overstocking by cattle alone can cause soil 'poaching' or 'puddling', which churns the wet soil into a structureless plastic morass. Heavy machinery is even more damaging. Wheels compress and destroy the soil structure, and as a result create an impermeable surface which cuts down water infiltration and root development. Surface impact of heavy machinery has increased, because the need to maximize the use of expensive equipment, together with the development of contract-farming, has forced the farmer to cultivate his soils either when they are too dry or too wet, with often disastrous results. The problem then is as much one of untimeliness as of method of operation.

Mechanical impact is not, however, confined to the surface; the deterioration of soil structure by compression, deformation and smearing occurs throughout the whole depth of continuously ploughed soils. In the latter, structural breakdown is often accompanied by the creation of a *plough-pan* which forms where the ploughshare 'smears' over the sub-soil. Smearing and compaction, in time, inhibit water percolation; sub-soil drainage is impeded, and in the heavier poor-structured plough-layer above, the risk of saturation increases and yields inevitably begin to decrease. In the case of 'minimum cultivation' it has been suggested that surface compaction by heavy drilling and harvesting machinery alone is sufficient to cut down infiltration rates sufficiently to counteract any advantages derived from lack of surface and soil disturbance.

Soil structure—the key to soil conservation

Therein lies the fundamental dilemma with which man has been faced ever since he first began consciously to exploit the soil, either directly for crops or indirectly for animal products. How does he obtain maximum yield without depleting the capital resource? It has become increasingly obvious that of all the completely inter-related variables over which he can extend some measure of control, none is so critical as the physical condition of the soil. The output of plant and animal products can be, indeed has been, spectacularly increased in many parts of the world by feeding, watering and protecting specially-bred potentially-high-yielding crops and stock. However, maximum return on these 'inputs' will not be achieved or (what is more important) sustained, unless the soil is maintained in as good condition as possible. Indeed it would hardly be an exaggeration to suggest that, within a given set of climatic and pedological factors, *soil structure* is the most important limiting factor as far as optimum yield is concerned; in view of the growing absolute shortage of certain artificial fertilizers, it must become even more critical to the efficient use of those available.

Soil structure, then, is the key to soil conservation in the widest sense. Man has still some way to go before solving the dilemma of directing his management in terms of both technical expertise and cropping systems to this end. It necessitates a greater recognition and reconciliation of short-term economic with long-term ecological aims if he is to minimize the undesirable effects of cultivation on the physical environment.

FURTHER READING

Brady, N. C. (ed.) (1973), *Agriculture and the Quality of our Environment*, Amer. Ass. Adv. Sci. Publication 85, Washington D.C.

Butzer, K. W. (1974), 'Accelerated Soil Erosion: a Problem of Man-Land Relationships', pp. 57–74 in *Perspectives on Environment*, ed. by Ian R. Manners and Marvin W. M. Kesell, Association of American Geographers, Publ. No. 13.

Cantor, L. M. (1967), *A World Geography of Irrigation*, Oliver and Boyd, Edinburgh.

Chorley, R. J. (ed.) (1969), *Water, Earth and Man*, Methuen, London.

Detwyler, T. R. (1971), *Man's Impact on Environment*, McGraw-Hill Inc.

Davies, D. B., Eagle, D. J. and Finney, J. B. (1972), *Soil Management*, Farming Press Ltd., Ipswich, Suffolk.

Ellison, W. D. (1944), 'Studies of Raindrop Erosion', *Agric. Engineering*, **25**: 131–136 and 181–182.

Ellison, W. D. (1948), 'Erosion by Raindrop', *Scientific American*, November, No. 817.

F. A. O. (1960), 'Soil Erosion by Wind and Measures for its Control on Agricultural Land', *Agric. Development Paper*, 71, H.M.S.O., London.

F.A.O. (1965), 'Soil Erosion by Water: Some Measures for its Control on Agricultural Lands', *Agric. Development Paper*, 81, H.M.S.O., London.

Grigg, D. B. (1974), *The Agricultural Systems of the World. An evolutionary approach*, Cambridge University Press.

Gourou, P. (1974), *The Tropical World*, 4th ed. Translated by E. D. Laborde, Longman, London.

Hudson, N. W. (1973), *Soil Conservation*, Batsford, London.

Ministry of Agriculture, Food and Fisheries (1972), *Modern Farming and the Soil*. Report of the Agricultural Advisory Council on Soil Structure and Soil Fertility.

Nye, P. H. and Greenland, D. J. (1960), *The Soil under Shifting Agriculture*. Technical Communication No. 51, Commonwealth Bureau of Soils, Harpenden.

Pereira, H. C. (1973), *Land Use and Water Resources*, Cambridge University Press.

Pilpel, N. (1970), 'Crumb Formation in the Soil', *New Scientist*, 31 Dec., pp. 582–585.

Tivy, J. (1971), *Biogeography: A Study of Plants in the Ecosphere*, Oliver and Boyd, Edinburgh.

CHAPTER THREE

USING PLANTS OPTIMALLY

N. W. Pirie

Introduction

Plants supply, or can supply, food in three ways: they can be eaten after no more pretreatment than is traditional in the kitchen; they can be processed to make them edible, or edible components can be extracted from them; they can be converted into more desirable or assimilable foods by conversion in animals, bacteria and other organisms unable to photosynthesize. The boundaries between these categories are not fixed. During the past century, especially in industrialized countries, the middle category has tended to absorb aspects of the other two. For example, cereal grains are less often made into flour or meal at home, and crops such as soya, groundnut and rape, which traditionally supplied edible oil and a residue fit only for use as cattle fodder, are now being processed more hygienically and completely into human food. There is nothing basically new in the idea that a crop is something that should be separated into fractions with different merits; it is as old as winnowing grain, extracting cane sugar, pressing out oil, or washing the cyanogenetic glucosides from manioc. All that is new is increased technical capacity to handle less tractable material.

The health of a community depends both on the evenness with which food is distributed within it, and on the total amount of food eaten. It is difficult to obtain accurate evidence about distribution, because people who are being observed are likely to change their food habits during the period of observation, and people who are questioned are likely to report what they think they should be eating rather than what they actually eat. The simple question, 'What did you eat at your last meal?' probably elicits the most reliable answers. It is reasonable to conclude that about 15% of a population will be malnourished, according to theory and the criteria of malnourishment used, when the amount of food available is 20% greater than the presumed requirement; and that twice the requirement must be available to ensure that only 5% of a population will be malnourished. There is, furthermore, an unfortunate discord between need for food and access to it. The young and the ill may get less than their share, and those

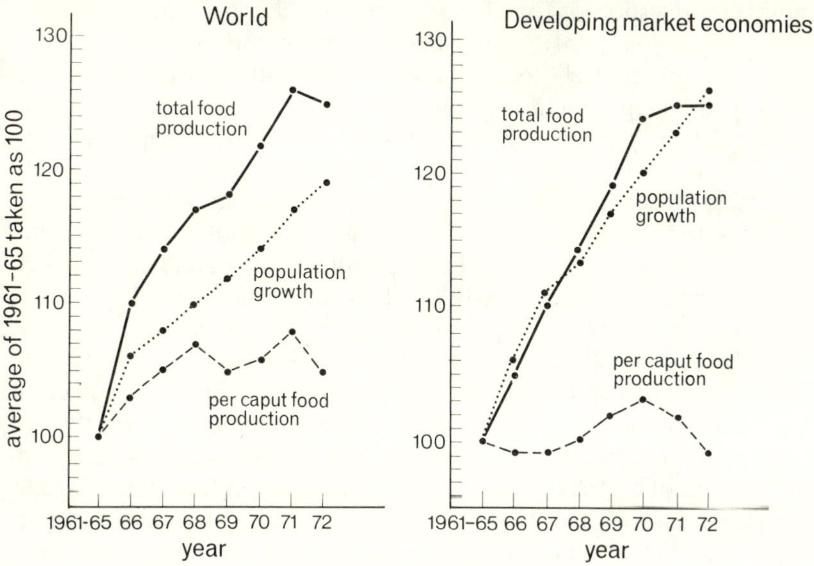

Figure 3.1 Food production and population growth 1966 72.

with intestinal parasites may absorb food inadequately, though need is enhanced. Factors such as these should be borne in mind whenever a nutritional assessment is made by simply dividing the quantity of food in a region by the number of people who inhabit it. In spite of these uncertainties, and of the unevenness with which countries supply FAO with reliable statistics, it is generally agreed that there is a food shortage now, that the position (as figure 3.1 shows) is getting worse, and that it will get still worse as the world population increases.

Protein supply

Protein supplies cause most anxiety because protein deficiency is more widespread than energy deficiency. Protein deficiency will not be abolished simply by increasing the scale of conventional arable farming; this is primarily a means for making carbohydrate. Statements such as these gained general assent a few years ago. It could be argued that assent did not lead to adequate practical action, and that discussion was not helped by the use of the foolish phrase 'protein-calorie malnutrition'. People living on diets that are deficient in both protein and calories are simply hungry. Logically, the more elaborate phrase means that, in spite of hunger, they get an adequate supply of vitamins, minerals and perhaps fibre.

The proposition that protein shortage is our most serious problem is now being questioned. This change of opinion arises partly from the usual swing of fashion, partly from a belated recognition of the well-known fact that protein-rich foods are used wastefully by people getting inadequate energy, but mainly from some basic misconceptions about the validity of the guessed values for human protein needs announced by various national and international committees. Nevertheless, it is likely that scarcity of protein is the most serious nutritional problem in the less-developed countries, that it will not be abolished simply by increasing the scale of conventional agriculture, and that research as well as development will be needed before diets can be improved. That is why, in this article, emphasis is put on the newer plant foods (some not yet commercially available) that are rich in protein, especially if they are adapted to small-scale production by simple techniques, rather than on foods that are primarily sources of fat or carbohydrate.

The precise meaning to be attached to such phrases as 'protein-rich food' or 'protein concentrate' depends on the amount of protein that is thought advisable in a diet. Many officials of national and international organizations argue that a diet is adequate if 10% of its dry matter is protein—but they themselves eat diets containing more protein than that. Their opinion depends on the fallacious use of nitrogen-balance measurements. These merely show how much protein is needed to prevent the body from being steadily depleted of nitrogen, which even the most ungenerous officials realize is a condition which could not persist for long, and that is probably harmful even when brief. These measurements tell us nothing about the composition of an optimal diet. Our nearest relatives, the primates, are healthier and more responsive now that most zoos give them 25% protein. It may be concluded, therefore, that an adequate human diet should contain more than 10% protein. Because some protein-free foodstuffs, e.g. fat and sugar, will be added to make the diet palatable, some of its components will have to contain 20% or more protein. It is foods of that type that are most urgently needed.

Leaves

The whole of our food supply depends ultimately on photosynthesis—even when pigs are fed on yeast grown on petroleum, for petroleum is a product of past photosynthesis. It seems logical, therefore, to start a survey of edible plant products with leaves—the dominant site of photosynthesis. Furthermore, in many (or possibly most) climates, leafy forages are the crop that gives the greatest annual yield per hectare. Proteins and other substances are synthesized in the leaf; there is therefore less waste of time and material if it is the leaf, rather than a seed or tuber which is filled by translocation from the leaf, that is harvested. According to current theory, the visible component of sunlight can be used for photosynthesis with 18% efficiency. Very good farming attains 3% and ordinary farming less than 1%. In laboratory conditions, with full protection from predation and disease, and with an optimal supply of fertilizer, water and CO_2, it is possible to reach 10%. The efficiency is greater in dim light, because many species cannot use full sunlight effectively, but the yield is obviously smaller. Figures such as these suggest that it should be possible to get annual dry-matter yields as great as 150 t ha^{-1} by using species adapted to high temperature and strong light. There seems to be no reliable evidence for yields more than half as large as that.

C3 and C4 plants

Two basically different biochemical processes are used in the first stages of photosynthesis by the crop plants that have been intensively studied. Crops developed in the Temperate Zone use strong light less efficiently, and transpire water more extravagantly, than many tropical plants. These Temperate-Zone plants are labelled C3 because, in them, CO_2 combines with pentose in an early phase of photosynthesis, and the resulting hexose splits to give two trioses. The mainly Tropical-Zone C4 plants combine CO_2 with a 3-carbon acid to form a 4-carbon acid. Figure 3.2 gives a somewhat idealized impression of the response of the more efficient members of the two categories of plant, in otherwise ideal circumstances, to differing intensities of illumination. Although the biochemical difference between C3 and C4 plants seems so fundamental, members of the two groups often appear in the same genus. They may even be able to hybridize. The moral is clear: when plants are being selected for use as fodder or as vegetables in regions where light is strong, special attention should be given to C4 plants even if, in their present uncultivated state, they seem less promising than C3 plants. The latter are usually the end product of many generations of selection for use in the Temperate Zone; a comparable amount of selection should bring about remarkable improvements in those plants that are already adapted to the tropics.

The yield of leaves

The simplest way to use the productivity of leaf crops is to eat leafy vegetables. Edible species, harvested when they are so immature that an

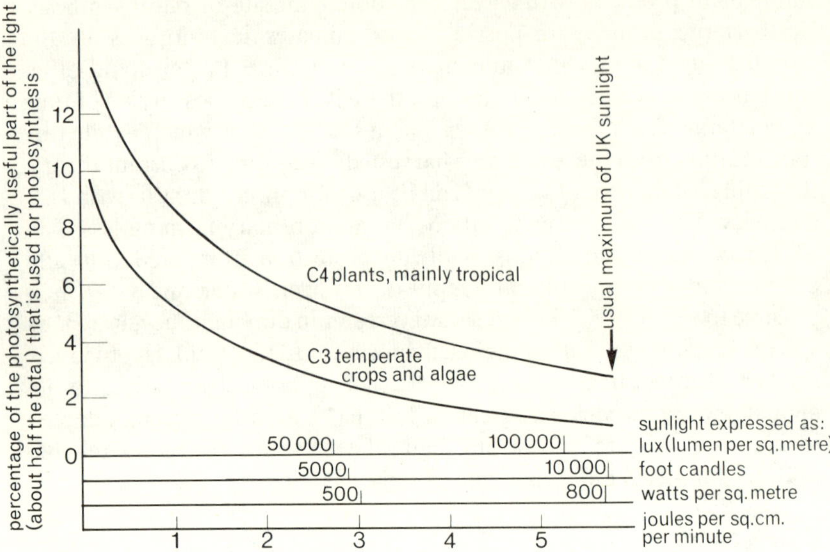

Figure 3.2 Response of the more efficient members of the C3 and C4 plant groups to different intensities of illumination.

unacceptable amount of fibre has not been formed, are not as productive as forage species, and only part of the whole crop is edible. Nevertheless, leafy vegetables produce more edible protein than any other conventional method of husbandry. In Britain a good crop of brussels sprouts (*Brassica oleracea*) yields 0·5 t ha^{-1} of edible protein in 6 months; an outstanding crop can yield 1 t. In the tropics, annual yields of 4–6 t ha^{-1} of edible protein have been reported from several institutes. Unfortunately, most publications dealing with horticulture or market gardening do not state the growing period, or whether the weight recorded is 'as harvested', 'as sold', or 'as eaten'. They sometimes do not even state whether it is the wet or the dry weight. The nitrogen content is seldom recorded, and the percentage of that N that is protein-N even less often. This is a serious defect in publication, because age at harvest, and cultural conditions, affect the protein and dry-matter content of leafy vegetables much more than they affect the composition of seeds or the underground parts of plants. The first step in any attempt to get market gardening taken seriously as a means of producing food is to publish experimental results in the complete form usual in other branches of agronomy.

Leaves as protein sources

Leafy vegetables are usually esteemed for their appearance, or for their vitamin and mineral content. Though important, these merits (the presence of β carotene—pro-vitamin A—is specially important) should not overshadow the importance of leafy vegetables as protein sources. Those eaten extensively in various parts of the world contain 3–7% nitrogen. The non-protein-N would be lost in conventional cooking, but most of the N is protein-N, and that is retained. Some communities eat 200–500 g (wet weight) of leafy material daily; 50–100 g is more usual, even in countries where vegetables are widely used. Table 3.1 contrasts the extent to which commercially grown vegetables, including non-leafy market-garden products, contribute to the protein intake in different countries. Bearing in mind that Table 3.1 does not include produce from private gardens,

Table 3.1 Grams of protein, per head per day, supplied by commercially grown vegetables (From FAO *Production Yearbook* 1971, Table 137)

Portugal	7·8	Ceylon	2·0
Italy	5·1	Denmark	1·7
Japan	5·1	Nigeria	1·1
France	5·0	Guatemala	1·0
USA	3·8	Brazil	0·5
Israel	3·7	Mexico	0·4
Chile	3·3	Venezuela	0·2
UK	2·6	India	0·1

it is reasonable to conclude that leafy vegetables are a protein source of about the same importance as fish. They get much less attention.

Leaf fibre

The protein-rich plants that can be eaten after no more processing than can be managed in the kitchen are not as productive as the forages. The species used as forage are, however, often ill-flavoured or are thought to be too fibrous to be used as human food. Opinion on the second point may be about to change. Various types of gastro-intestinal disease seem to be more common in industrialized than in developing countries. The idea that this is a consequence of the relative absence of fibre from the food eaten in industrialized countries is gaining acceptance. If the idea should be substantiated, it would be reasonable to reinvestigate the use as human food of some forages, e.g. oats harvested at the 10–15 cm stage when the leaf contains more than 25% protein. In spite of this possibility, there are limits to the capacity of the human digestion to cope with fibre; this is especially true in infant feeding.

Extraction of protein from leaves

There is therefore increasing interest in processes for extracting edible protein from forages and other leafy material. The fresh crop is pulped,

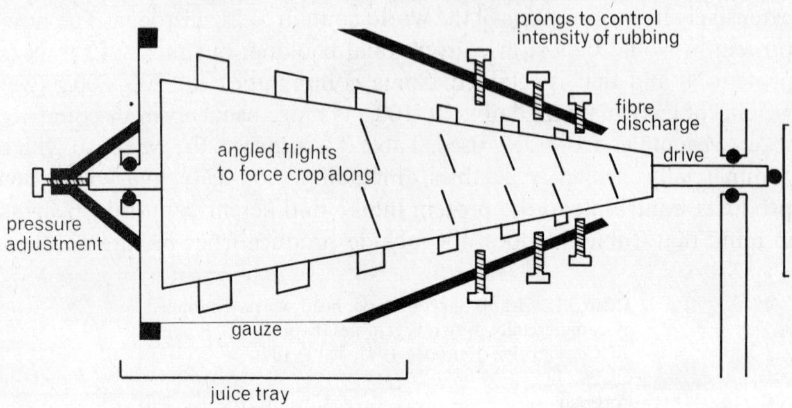

Figure 3.3 This suggested design for a leaf protein extractor differs from conventional screw expellers in that the scroll is broken to such an extent that it is primarily a means of rubbing and bruising the crop rather than simply pressing it. Furthermore, the extract runs out, rather than being pressed out, thus diminishing the amount of fibre that is carried along with it.

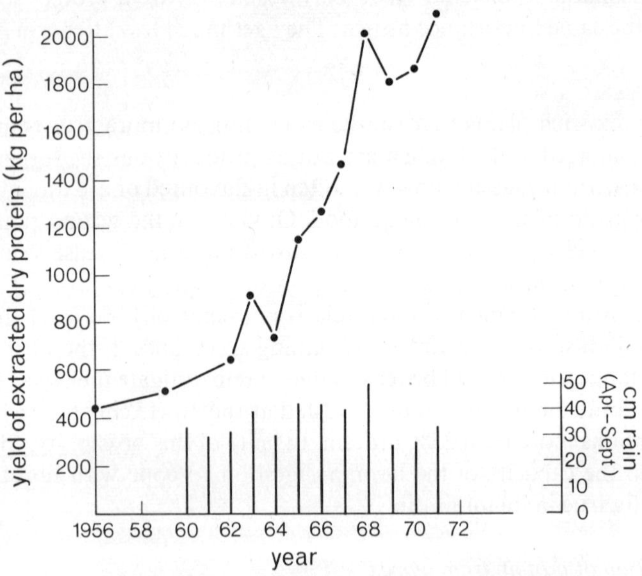

Figure 3.4 Improvements in the annual yields of extracted leaf protein.

protein-containing juice is pressed out, and the protein is coagulated and filtered off as an edible mass that is easily incorporated into various types of food. Machinery for extracting protein is simple and is continually being improved. A new design, still in the experimental stage, is shown in figure 3.3.

The case for growing a crop primarily as a source of extracted leaf protein depends on the observation that the yield of protein, in the form of an edible concentrate, is greater by this form of agriculture than by any other. Figure 3.4 shows how yields increased as we gained skill. The limit in Britain is probably 3 t ha^{-1} of dry extracted protein per year. In India, with no winter cessation of growth, that yield has already been reached; it should be possible to get 5 t ha^{-1}.

The percentage of protein that is extracted depends on the species used, and the age and protein content of the leaf at harvest; it is usually 40–65% of the total protein. Unextracted protein makes the residual fibre a useful ruminant feed. The residue can be dried to produce conserved fodder much more cheaply because the fibre can be pressed till a ton of dry matter is accompanied by little more than a ton of water, whereas the ratio is 6 or more times as great in the original crop. In Britain and the United States, more importance is attached to the economically dried forage, for use as winter feed, than is attached to the extracted protein.

The plant species and varieties now used as sources of seeds and tubers are the end products of prolonged selection. They will no doubt be improved, but much of the basic work has been done. There has been no similar attempt to select plants that are well adapted to protein extraction. For example, much of the work in Britain is done on cereal varieties selected for their efficiency in forming seed, but they are harvested for protein extraction when immature. Varieties rejected by plant breeders because of their poor performance as seed yielders, may well be preferable as yielders of leaf.

In the British climate there is probably little need to concentrate attention on species or varieties able to use strong light efficiently, i.e. C4 rather than C3 plants. The important qualities in plants that are to be used as sources of leaf protein are the ability to start growth early in the year, and the ability to cover the ground quickly, so as to intercept as much light as possible. Perennial plants and annuals that can survive the winter are therefore likely to give good yields, but some of the weeds that start to grow earlier in the year than conventional crop plants, e.g. jack-by-the-hedge (*Sisymbrium alliaria*) and bluebells (*Endymion non-scriptum*) may be worth examining. In any climate, the ideal plant will have leaves that do not mature in such a manner that the protein becomes less extractable (this is probably usually the result of the accumulation of phenolic compounds), and it will be able to regrow after harvest. There will be little support for research on points such as these until agricultural administrators have grasped the fundamental proposition: forage crops do not have to be used unfractionated; they cover the ground more completely and continuously than other crops; consequently, in many climates they give better yields and better protection from erosion.

Leaves produced as a byproduct and at present wasted are obviously an attractive source of leaf protein. Their potentialities have not yet been fully studied. Sugar beet tops yielded 800 kg ha^{-1} of protein; haulm from peas harvested for freezing or canning yielded 600, which is considerably more protein than is in the peas. Potato haulm harvested at the beginning of September, when it is usually killed with acid as a protection against blight, and to facilitate lifting the tubers, yielded 300 kg of protein and 3 t (dry matter) of fibre containing as much N as good hay. More protein is extracted from haulm taken earlier—but tuber yield is diminished. Much of the solanin, which makes haulm too toxic to be used in quantity as fodder, is removed in the process; if enough remains to be hazardous, it could be removed by re-extracting the protein and fibre in slightly acid conditions. There are similar byproduct leaves in tropical countries. Tolerance of the waste of these resources is not just immoral—it is daft.

Stems

Stems and other similar parts are little used as food by people, though many animals eat bark. Sago (*Metroxylon*) is a source of starch, and sugar cane (*Saccharum officinarum*) of sucrose. So much cane juice is processed that more attention should be paid to the coagulum containing 25% protein which can be separated from it. The production of 1,000,000 t of sugar involves handling 'mud' containing about 10,000 t of protein and this, with modern techniques, is a potential human food.

Flowers and immature seeds

Flowers, in various stages of development, are important as spices and flavouring agents. In the form of broccoli, calabrese and cauliflower (*Brassica* sp.) they are useful vegetables because they contain 25 to 30% of protein, provided the thicker pieces are excluded from the sample analysed. Some people do not like the dark green colour of the more nutritious leafy vegetables and, because they have been told that they should eat vegetables, choose the much less nutritious pale inner leaves of cabbage. If pallor is important, they would be better advised to eat immature flowers.

The pods and immature seeds of various varieties of *Phaseolus*, e.g. runner beans, if not taken too early, contain more protein and digestible carbohydrate than the mature easily-conserved seeds. This is also true of soya (*Glycine max*) picked green. In climates where multiple cropping is possible, more crops can be grown in a year if crops are harvested when immature rather than being left to ripen. The annual yield of protein from four crops of green soya can be 3·2 t ha^{-1} (which is more than twice the protein yield from mature soya) and the product does not have the defects, e.g. a flavour that some people find unattractive, of the mature bean. Because of these merits, the immature reproductive parts of plants deserve more attention than they get; it is, however, unlikely that they will ever play as important a part in human nutrition as less perishable foods. Without agreeing with the more extreme claims for the medical and nutritional merits of fresh home-grown foods, it is worth pointing out that the distribution of perishable plant material raises no more problems than the distribution of fish or milk.

Although perishable fresh vegetables, and perishable protein concentrates from such sources as beans, coconuts and leaves, could with advantage be used more extensively than they are, some foods with a long shelf-life are needed. They are essential as stores to meet temporary crises such as flood and earthquake; it is energy-giving foods rather than protein concentrates that are then needed.

Countries with predictable periods of drought or cold need stored protein concentrates if they are to maintain diets of more-or-less constant composition throughout the year. And countries that have to import much of their food because they are densely populated, must usually include storeable protein concentrates among the imports. Food preservation and storage is therefore assiduously studied in all countries where research is undertaken. The conversion of foodstuffs into products with an indefinite shelf-life may indeed have become an obsession. Research seems to be organized by people who envisage the whole of nutrition in terms of neat packets on the shelves of a supermarket. As a result, too little attention is given to making and distributing foods with a short shelf-life. They therefore tend to be expensive.

Seeds

The outstanding merit of mature seeds is that they can be stored. But conditions of storage are so bad in many countries that rats are said to eat 30 Mt* of grain a year, and to damage more. Fungus infection in field and store causes loss and may make the remains of the seed toxic. Insects cause loss by transmitting infection and by predation in field and store. Alarm is widespread at the use of toxic chemicals to control these harmful organisms; sober assessment suggests that the alarm is based more on emotion than fact. Nevertheless there is risk from toxic chemicals, even

* Mt = megatonne = a million ton(ne)s.

when carefully handled; this stimulates research on less toxic chemicals and on biological methods of control. At present we have to compare the risk from food shortage, mycotoxins, etc., with the risk from pesticide residues. It seems certain that the former is the larger—more people are being malnourished than poisoned.

Cereal seeds

Cereal seeds supply about a third of the protein eaten in Britain and, because the commonly-used varieties contain 10–12% protein and that is the average percentage of protein in the diet, they also supply about a third of the energy. They supply a larger proportion of the energy in less-well-fed countries. Annual production in the world exceeds 1000 Mt, but that figure cannot be used directly to assess the amount available as human food, because nearly half is used as fodder. There have been several recent developments of great scientific and practical interest.

It was clear 30 years ago that most of the photosynthesis which filled a wheat (*Triticum aestivum*) grain took place after the ear had formed and in its vicinity; however useful the straw and lower leaves of a fine tall plant may have been in primitive conditions to help the plant to compete with weeds for sunlight, they wasted time and material when farming became efficient. Short-strawed cereals are the basis of the 'Green Revolution'. They are often called *high yield*, but would be better called *high response*; their merit is that, partly because of the short straw, it is worth while giving them quantities of fertilizer to which traditional wheats would not have responded, or that would have made them top heavy if they had responded. A similar appreciation of the principles of plant physiology has directed the attention of maize (*Zea mays*) breeders away from strains carrying a few enormous cobs and towards strains with many smaller ones. This is because photosynthesis is often limited by the inability of the leaf to get rid of the photosynthate into a 'sink' quickly. Many 'sinks' make that easier. Cereal breeders are also tackling the problem of protein deficiency. It has long been known that individual grains in an ear of wheat could range in protein content from 7 to 21%—the grain therefore has no intrinsic inability to lay down protein. Strains of barley, maize, sorghum and wheat that contain more than 20% protein on bulk samples have now been developed and are being tested for yield, disease resistance, and suitability for making foods such as bread, chapatties, pasta and tortillas, in which texture is often deemed as important as nutritional value. An unexpected consequence of these developments is increasing interest in oats (*Avena sativa*); hitherto a crop despised by progressive farmers. There are strains containing 30% protein which will grow in worse climates than the other cereals. By suitable milling techniques, protein-rich fractions can be made from oats that contain less protein initially.

Protein concentrates are made from low-protein cereals by mechanical fractionation. The protein-depleted portion is not wasted; it can be used as a substrate for growing micro-organisms or, with supplementary urea or some other nitrogen source, as ruminant feed. The method of fractionation must be adapted to the manner in which the protein is distributed within the grain. In rice (*Oryza sativa*) protein is concentrated in a layer immediately under the bran. It can be separated from the main body of the endosperm by suitably contrived abrasion; equipment for doing this is used in various parts of the world. Starting with rice containing only 8 or 9% protein, fractions containing up to 21% protein can be separated. It is claimed that by combining abrasion with solvent extraction, the yield of the protein-rich fraction can be increased. This process also concentrates the thiamine and riboflavin. These products are already being used commercially in baby foods. In a somewhat similar manner, wheat flour is separated in an air stream into protein-rich and protein-depleted fractions. This is more convenient than the traditional Chinese wet process for making gluten (*mien chin*) which was introduced into Europe by Beccari in 1728. Fractions containing 30% protein can be made. Protein-enriched wheat flour is used regularly for making starch-reduced bread.

Legume seeds

Legume seeds are the traditional protein concentrate in much of the tropics and sub-tropics. They have not hitherto had as much attention from scientists as they deserve, nor as much encouragement from departments of agriculture. In India, annual production was about 12 Mt for many years; there was then a slight diminution because of increased use of land for 'high-response' cereals, but there is now a welcome increase in interest. The conclusions from research by the Indian Council of Agricultural Research cannot be summarized briefly because species need different times to mature, and the possibility of harvesting 2 or 3 crops a year depends on the distribution of rainfall and the availability of irrigation water. Furthermore the optimal time of planting, spacing, and use of fertilizers still needs fuller study. In the ideal conditions of an Experiment Station, yields of seed can reach 4 t ha^{-1}, and the rate at which protein is synthesized can reach 4·7 kg ha^{-1} per day. The seeds of legumes that grow well in Britain are usually eaten when green and immature. An exception is the field bean (*Vicia faba*) which contains 25% protein and is more prolific than the broad bean, a large-seeded variety usually eaten when immature. Field bean is being considered as a source of isolated protein for use in human food.

In good conditions, edible legume seeds can supply about 1 t ha^{-1} of protein per year as a concentrate containing 25 to 30% of protein. The nature of the other material in the seed is not always known, and it may sometimes limit use. Thus about half the weight of shelled groundnuts is useful oil, whereas some beans contain useless polysaccharides; some contain oligosaccharides that are fermented in the gut and produce unacceptable amounts of gas, and some take so long to cook that they are not welcome, although they give good yields and have pleasant flavours. Advocacy of groundnuts (*Arachis hypogaea*) is sometimes condemned because of the risk of aflatoxin poisoning. This is hysterical. Any food can support the growth of funguses such as *Aspergillus flavus* (which produces aflatoxin) if conditions of storage are primitive. There is reason to think that other mycotoxins cause liver tumours in several countries. The association with groundnuts arises because, coming from the ground, they are more liable to contamination than other crops; they are often stored in worse conditions, and inferior material is often fed in bulk to animals. Animals have certainly been injured by aflatoxin; people are more discriminating in what they eat; perhaps that is why there is as yet no evidence that they have been injured. Groundnut meal from which the oil has been expressed contains 50% protein and is coming into use as a protein supplement in infant feeding, especially in India. Groundnuts cooked in various ways are traditional foods in many parts of the world. It is therefore surprising that so much effort is now being directed towards producing protein isolates and dried solvent-extracted powders rather than towards peanut butter which is already used extensively in the United States.

Cereal seeds have little flavour and, when handled in traditional ways, have few harmful properties. Bran is easily removed during milling, although many authorities maintain that diets would be improved if it were not removed; and phytic acid, which interferes with calcium absorption, is largely destroyed when flour or meal is kept warm and moist for a few hours, as in traditional methods for making bread and porridge. Legume seeds contain more troublesome deleterious components. Because of its resistance to drought, *Lathyrus odoratus* is grown in parts of India. There are several toxic substances in some strains, but they appear to be removed if the peas are boiled in a large volume of water which is discarded. Most legume seeds contain trypsin inhibitors and haemagglutinins; these are destroyed by thorough cooking. Extensive experimentation in normal animals shows that raw beans diminish growth rate and cause pancreatic enlargement; inadequately cooked beans are thought to be harmful if eaten in quantity by people. The trypsin inhibitor from soya has little effect on human trypsin; and those raw legumes that have been tested are not toxic to germ-free animals, though they become toxic when some members of the normal gut flora are introduced. The reasons for the beneficial effect of thorough cooking are therefore obscure. Many reasons for the harmful effects of raw beans have been suggested, e.g. the presence of a heat-labile substance that becomes toxic as a result of microbial attack (such a substance is a known component of *Cycas* seeds); the passage of unusually large amounts of protein into the lower gut where it undergoes harmful microbial attack; loss of S-containing amino acids in secretions from the hypertrophic pancreas; damage to the gut wall that permits invasion of the body by bacteria. It may be that the problem, here as elsewhere, is complicated because, in different circumstances, several of these harmful processes may operate simultaneously.

Thorough cooking inactivates the agents discussed in the last paragraph. Some legume seeds contain thermostable deleterious substances such as oligosaccharides causing flatulence and, in broad beans, an unknown component causing haemolytic anaemia (favism) in people congenitally deficient in glucose-6-phosphate dehydrogenase. Clearly, though very valuable, legume seeds must be used circumspectly.

World production of soya beans in 1973 was 58 Mt—about equal to the total of all the other legume seeds. About 4 Mt are eaten in East Asia where it is a traditional human food. Soya was introduced into the United States at the beginning of this century as a green fodder. Its value as a source of oil was then recognized, but much of the residue was (and sometimes still is) used as fertilizer or burnt. Its value as cheap cattle fodder was then recognized. The more perceptive observers already knew that byproducts remain cheap only so long as few recognize their value. Thus Lawes wrote in 1885: '. . . as I pointed out many years ago—linseed, cotton, and rape cakes are waste products; they would be manufactured just the same whether farmers purchased them or not, and the price at which they are sold is neither more nor less than what the competition between farmer and farmer enables the maker to obtain'. Now that byproducts are being valued as potential sources of human food, what used to be waste has a commercial value as great as, or even greater than, the product that was once primary.

Traditionally, soya was made edible by various types of fermentation, and by extracting much of the protein from the ground beans with water and coagulating it with gypsum. One reason why soya is rarely acceptable when cooked like other beans is thought to be the formation of bitter ethylvinyl ketone when lipoxidase has the opportunity of acting. The development of undesired flavour can be prevented by 'toasting' the beans or by cooking them rapidly. Methods of presentation based on fermentation and skilled cooking deserve much more attention than they get because they are well adapted to use on a small scale by people with little equipment. It is a pity that so much research attention is diverted to methods for making 'texturized vegetable protein' and 'simulated meat'. This work is no doubt profitable, but it is not likely to have much effect on the diets of those now most in need of more food.

Valuable though soya is, it is not suited to all climates. Skilled research is extending its climatic range. This is done partly because of the prestige that soya has acquired, and partly because the present varieties are the result of half a century of breeding and selection. Given the same amount of scientific attention, it seems likely that several other legumes, already climatically adapted, would surpass soya in some climates. One reason for suggesting the intensive study of other legumes is invalid; it is only because soya is already used so extensively that its potentially harmful components have been found. There is no reason to think that it raises more problems than other legumes will raise when they have been as thoroughly studied.

Cotton seeds

Cotton (*Gossypium hirsutum*) is pre-eminent among seeds from plants other than cereals and legumes. Because of the presence of glands containing brown pigments and the phenolic aldehyde gossypol, the seed was for many years used simply as cattle fodder. There are strains free from these glands, but they are more susceptible to insect attack. By judicious milling, followed by air classification or liquid separation, the glands can be removed, so that a meal is made with a more attractive appearance. It also has better nutritive value, because gossypol makes part of the lysine in the seed protein unavailable. Cottonseed meal containing 50–60% protein has been used on a small scale in nutrition programs and is on sale in Central America; it is now likely to become more important elsewhere. Similarly rape (*Brassica napus*) was for long merely a minor source of oil. Because of the presence of erucic acid in it, the oil could not be used extensively, and the meal contained toxic thioglycosides. In spite of these initial disadvantages, rape seems attractive as a non-tropical oil source; strains have been selected containing little erucic acid, and the other toxic components have been removed by breeding and selection. It is not yet used extensively, but it is likely soon to become an important source of food.

Other seeds

In different places, and at different times, many other seeds have been sources of oil and protein. It is impossible to arrange them in order of importance, because the position is fluid as a result of changes in processing technology and in the changing nutritional and industrial merits attributed to the oils. Sunflower (*Helianthus annuus*) is one of the richest in edible oil, whereas safflower (*Carthamus tinctorius*) oil is highly prized because of the presence of unsaturated fatty acids, and castor (*Ricinus communis*) is an important lubricant. All yield residues after oil extraction contain 40–60% protein, depending on the completeness with which the kernel is separated from the hull or pericarp before the seed is ground for oil extraction. Problems are associated with the use of each residue. Sunflower meal is apt to be discoloured because of the presence of chlorogenic acid and other phenolic substances; safflower meal is bitter, and the allergens in castor meal must be carefully detoxified. Processes for doing these things are known and are continually being improved.

Partly because they are traditional, and partly because they suit the local climate, many other seeds deserve attention. Cowpea (*Vigna unguiculata*) and winged bean (*Psophocarpus palustris*) are important in parts of West Africa and Southeast Asia, especially in regions ill-adapted to soya. Several thousand years ago the seeds of fat hen (*Chenopodium album*) were eaten in Europe, those of *C. ambrosioides* and *C. nutalliae* are eaten now in Mexico, and *C. quinoa* seeds are made into flour and eaten in the highlands of South America. Although quinoa has not been subjected to up-to-date selection, and is probably suboptimally fertilized, it yields up to 3 t ha^{-1} annually. The seed contains 15% protein. A more broad-minded approach to the choice of plants as sources of edible seeds would probably be beneficial.

Seeds and other reproductive parts of trees have a varied role in human nutrition. Fruits are important sources of ascorbic acid, but probably supply little protein and about as much energy as is supplied by grapes in the form of wine. Bananas and plantains are important sources of energy in the wet tropics, but their dry matter contains only 4–5% protein. It is claimed that the annual yield from established groves of leguminous trees such as algaroba and carob can reach 50 t ha^{-1}. Coconut (*Cocos nucifera*) is the most important protein-yielding tree. There is, however, so much fibre in the 'meat' of the coconut that it can be eaten in small quantities only. Traditionally it is grated with warm water, and the resulting emulsion of protein and oil is separated from fibre by pressing through a cloth. The possibilities of using this method on a large scale are now being investigated, because the protein has good nutritive value. By conventional methods of processing, most of the protein can be used only as cattle fodder, because of the damage done to it when it is dried to make copra, and then over-heated in the course of expressing oil. The potentialities of trees as sources of various types of food would repay serious investigation. When tropical rain forest is cleared to make way for conventional arable farming, there is often serious erosion in spite of careful terracing and contour ploughing. The obvious answer is to try slowly to replace trees that do not yield food by trees that do.

Underground parts of plants

Cassava (*Manihot esculenta*), potatoes (*Solanum*), sweet potatoes (*Ipomoea*) and yams (of both the *Dioscorea* and *Araceae* groups) supply about 150 Mt of edible dry matter annually, i.e. about a fifth as much as is supplied by wheat and rice together. But they supply less than 10 Mt of protein, and most of this comes from potatoes. It is unfortunate that species adapted to growth in the wet tropics put so little protein into their underground parts, because dependence on cassava and yams is responsible for much protein deficiency. Better varieties could probably be selected; little effort seems to have been put into this research so far.

New crop plants

Traditions connected with the use of plants, and parts of plants, have a long history. We have moved plants from continent to continent, so that the traditional food of one community becomes for a time the non-traditional food of another; but 99% of our plant food comes from species domesticated 5000–10,000 years ago. It would be a useful break with tradition if we stopped assuming that primitive women had unsurpassable culinary wisdom, and started to look for potentially valuable species among plants now classed as useless weeds. There are probably many species that would respond to breeding and selection by up-to-date methods. The discovery of completely new domesticable species will be a slow and uncertain process. The principle has already been stated that more attention should be given to improving species that are already adapted to a climate rather than to adapting species that have already been greatly improved. It may be restated as: *find uses for what already grows well.* An obvious starting point is species that are already used, although they have not had the benefit of skilled selection.

Agave

A candidate species that is more unconventional than the improvement of C4 plants as vegetables or sources of leaf protein, and the improvement of food-yielding trees, is the maguey (*Agave atrovirens* and *A. americana*) from which pulque is made. In Mexico it grows with little attention, and on rough ground. After 10 to 12 years, juice is collected during 4 months and the plant dies. The juice from a hectare contains 70 t of mixed sugars, i.e. the annual yield during a cycle in which $\frac{1}{12}$ of an area was harvested annually would be 6 t ha^{-1}. Maguey may not be a probable source of crystalline sucrose, but it seems to have potentialities as a source of molasses.

Nitrogen-fixing plants

Leguminous plants have the outstanding merit that bacteria in nodules on their roots can fix atmospheric nitrogen. They turn a normally inert substance into nitrogen compounds that the legume, and plants near it, can use. Industrial fixation of nitrogen is a heavy consumer of energy—mainly in the form of oil and natural gas. The price of N-fertilizers has therefore greatly increased recently, and so increased the attractiveness of legumes. Not all members of the order *Legumiales* carry N-fixing nodules, and many plants in other orders fix nitrogen by means of nodule bacteria and in other ways. Interest in these non-legume plants is increasing, partly for purely scientific reasons. Because they are often pioneers after volcanic eruptions or glacial retreat, they are responsible for much of the productivity of swamps in cold regions; and diversity in the species used as crop plants is a safeguard against new strains of pathogen.

No non-legume plant is as effective as the better legumes. Thus good varieties of lucerne (*Medicago sativa*) can fix nitrogen at an annual rate exceeding 300 kg ha^{-1}, whereas alder (*Alnus glutinosa*) and some tropical grasses (*Digitaria* and *Paspalum*) manage only 100; but little effort has so far been expended on selecting effective varieties of these non-legumes. Blue-green algae contribute usefully to the N-nutrition of rice in paddy fields, bacteria on the surfaces of some leaves fix nitrogen, and nodules are known in which a phycomycete rather than a bacterium is the fixer. This is interesting as the only example of fixation by a nucleated cell (i.e. by a eukaryote). The importance of these types of fixation is that they suggest the possibility of introducing the ability to bear effective nodules into non-legume plants, and even of introducing the ability to fix nitrogen into the cells of plants themselves. According to thermodynamic theory, fixation does not consume energy; according to biological experience it does. Fixation may therefore divert photosynthate from the formation of starch and other plant components. It may, on the other hand, present the plant with a sink into which photosynthate can go. As already mentioned, the absence of a sink can restrict photosynthesis.

Algae

During the past 40 years, scientists in many different parts of the world have advocated the use of unicellular green algae instead of agricultural and market-garden plants. It seems now to be agreed that, on a practical scale, green algae have no special merits. In comparable conditions they are no more efficient at photosynthesis than conventional plants; they have to be supplied with CO_2 which conventional plants collect from the air; and they often need more careful shielding from invasion by other organisms. Furthermore, the end product resembles a crude preparation of leaf protein; any doubts about the merits of leaf protein apply even more strongly to algae. In some circumstances, however, green algae are useful. The abundant natural growth of algae such as *Spirogyra*, which has a habit of growth that permits hand-harvesting, is used as a food in Thailand, as is the tiny angiosperm *Wolffia*. Because algae do not transpire, they use less water than conventional plants, and so may have a place in desert farming. They are already being cultivated in sunlit flumes carrying sewage. The oxygen produced by their photosynthesis helps to purify the sewage, and the final mixture of bacteria and algae is fed to chickens. The blue-green alga, *Spirulina*, grows naturally in saline and alkaline lakes; its entangled manner of growth makes collection easy, and it is a traditional food in several regions. In Mexico, about a ton a day is now being harvested commercially, and there are plans for deliberate cultivation. None of the objections to green algae apply here. The alkalinity of the medium ensures a supply of CO_2 and restricts the growth of other organisms, and the area covered by an alkaline lake would probably not otherwise be used for agriculture.

New farming methods

Town-dwellers sometimes refer to farmland as 'natural'; it is in fact a wholly artificial creation. There is no reason to think that any part of conventional farming technique has reached the limit of advantageous artificiality. It is, nevertheless, worth while examining some recent suggestions critically. Laboratory work many years ago suggested that the main benefit from ploughing was weed control. The advent of selective herbicides, coupled with a fuel shortage, will probably diminish the amount of ploughing. A similar line of thought leads to the suggestion that soil merely gives mechanical support to a plant and holds nutrients near to its roots. These jobs could be done by gravel, plastic granules, or even water if the plants were suspended from a gantry. There is no reason to think that any of these alternatives would be any cheaper than soil. The growth of many cultivated plants is stimulated by increased CO_2 and decreased O_2. When plants are grown in a greenhouse, because of the need for warmth, it is reasonable to give them extra CO_2. Conditions in which the extra yield would compensate for the cost of removing O_2 are unusual—space stations are a small-scale possibility. The subject deserves more attention, because a changed atmosphere will probably affect such aspects of plant physiology as seed-formation. Part of the effect of changing the atmosphere around a plant is thought to be a consequence of the inhibition of photorespiration. By selecting varieties naturally deficient in this mechanism, yields should be enhanced. There is some evidence that they are.

If artificial light could be made cheaply, the benefits would be enormous. There is, however, no immediate prospect of this, and the illumination of many square kilometres cannot be envisaged. Publicity is sometimes given to so-called 'plant factories' in which there is growth in superposed, illuminated trays. Where land is very expensive, such an arrangement might be advantageous; it is hard to judge because some of the published figures relate to conditions in which there is very little photosynthesis. What is called 'growth' is no more than the transference of stored material from seed into leaf.

With mechanized harvesting it is usually convenient, or even essential, to sow a field with a single species, carefully selected so that the seeds germinate simultaneously and the crop matures uniformly. In countries with much unemployment, harvesting by hand is advantageous, and uniformity less advantageous. Sequential ripening is convenient in a kitchen garden. There are some not wholly convincing claims that a mixture of species yields more than either species alone; predators and pathogens can obviously spread less easily in a mixed crop. The merits of mixed cropping deserve fuller investigation: perhaps we are unduly inhibited by the prohibition of maslin in Leviticus.

Saprophytic plants

All the species considered so far are able to photosynthesize: they use solar energy to bring about the reduction of atmospheric CO_2 to carbohydrates and other plant components. Mushrooms and similar large funguses are the only non-photosynthetic plants regularly eaten by people fully aware of what they are eating. Bacteria, yeasts and the smaller funguses are regularly eaten, incorporated into foodstuffs such as bread, soya or cassava that have been improved by fermentation, and in beers that have not been carefully filtered. In these products, the saprophyte is valued more as a means of bringing about a change than as a component in the ultimate food. The idea that some of the smaller saprophytes would be useful foods in their own right has been gaining ground, largely as a result of the enterprise shown in the mass cultivation of yeasts on hydrocarbons, and of bacteria on methane and on methanol derived from it. These substrates are non-renewable; they are becoming more expensive, and the technology involved in using them is intricate. Fermentation techniques in which the substrate is an abundant plant carbohydrate, or a byproduct, are more immediately relevant. With the addition of cheap sources of nitrogen and phosphorus, the saprophyte grows at the expense of part of the carbohydrate. The resulting mass contains mycelial protein as well as the residual protein of the seed. Thus an *Aspergillus*, growing on barley and urea, produces an end product containing 31% protein. This is a satisfactory pig feed. *Rhizopus* is similarly grown on cassava. *Paecilomyces* and *Candida* are grown on liquid substrates such as molasses and the effluent from paper-making; a *Rhodotorula* is grown on coconut water. There are obvious technical difficulties in separating small organisms such as these from the culture-fluid. One of the advantages *Spirulina* has over the green algae, discussed earlier, is that it forms a tangled mass that is easy to collect. The mycelium of *Fusarium* has an even more-convenient spongy texture. Unlike the yeasts, or proteins extracted from oil-seed residues, it already has a somewhat meat-like structure and does not need to be 'texturized' before being presented as a food.

Some forecasters suggest that fermentation along the lines discussed in the last paragraph, supplemented by fermentation of fibrous-plant by-products, will lead to the disappearance of animals as sources of food. This is very unlikely. The ruminant is a self-reproducing mobile fermentation unit with built-in temperature and pH-control. Its value as a consumer of vegetation growing on rough or steep areas of land is obvious. The relative merits of ruminants and fermentation tanks, bearing in mind the widespread preference for animal products, are a matter for experiment rather than assumption. However, it seems certain that ruminants will be preferable in non-industrialized countries. They are familiar; little mechanical or chemical skill is needed for tending them; and the small size of each 'fermentation unit' obviates the need to cart bulk fermentable material from the places where it is produced to a large central fermenter.

Conclusion

A reasonable social ideal is that each region should be self-sufficient with respect to the essentials of an adequate diet for its people. The agricultural ideal is that each region should produce the crops for which it is climatically best suited; the problem is to reconcile these often incompatible ideals. Wealthy countries have in the past imported food, thus contravening the first ideal; and they have contravened the second by growing crops that were thought for various reasons to be desirable, although climatically inappropriate. All countries have been wasteful. Water weeds that are a stinking nuisance in some tropical countries, are carefully collected and fed to animals in others. In Britain 81% of sugar beet (*Beta vulgaris*) tops are ploughed in, and thus used inefficiently, whereas in West Germany only 10% are ploughed in. As much effort is now being put into finding ways for destroying the dung from intensively housed cows, pigs and poultry, as is put into getting it onto the land as fertilizer. Energy is wasted when traditional routines are followed that may be in part unnecessary. Thus the annual cost of preparing the seedbed for cereals in Britain is £50 million. No one knows how much of the work done is necessary. Some of it may even be harmful.

Agriculture uses such a small part of the total fossil fuel used in a country such as Britain that forecasts about the imminent collapse of farming because of fuel shortage are alarmist. Fuel is, however, getting more expensive, and it is always well to be economical. Hence the stress put here on optimal use of crops. The need for economy is even greater in less-prosperous countries. Power is used on the farm to save human work, and it is likely that it will continue to save muscular work; but the use of power for driving machinery and synthesizing herbicides so as to eliminate jobs that are discriminatory rather than intensely muscular, e.g. weeding, may have been overdone—especially in countries with an unemployment problem. In future, more emphasis will be put on yield per hectare and joule, and less on yield per man-hour.

To get maximum agricultural productivity, many changes in crop plants and methods of using them will be needed. These changes are obstructed by lack of imagination, defeatism and illusion. In answering questions at a recent meeting a distinguished agricultural expert said there was no need for any new species of crop plant. A few minutes later he told another questioner that the idea of farming land that is now tropical rain forest was absurd, because of the risk of erosion if the land were ploughed. His imagination could not grasp the idea that one should find a perennial crop plant that would protect the ground from erosion. In this attitude he gets staunch support from defeatists who say it has been tried and has failed. As if that were not enough, self-styled ecologists spread the illusion that tropical rain forest must be preserved as a main source of oxygen. No region is a net producer of oxygen unless the amount of living or dead (e.g. peat) plant matter on it is steadily increasing. For as long as decay and consumption by animals maintains constant conditions, oxygen is being consumed as fast as it is formed. Farmland that produces a crop that is carted away and eaten elsewhere, produces more oxygen locally than forest; wind redresses the balance.

As progress is made towards optimal use of crops, there will have to be changes in eating habits. They may amount to no more than increased use of a familiar food, such as green vegetables, or to the use of foods that are familiar in other countries. But some of the products discussed here are not familiar anywhere. If they are to be used, extensive research on methods of presentation and on the techniques for winning acceptance will be needed. It is often argued that food habits are hard to change. A moment's reflection on changes in the pattern of food consumption in industrialized countries during the last half-century shows that this argument is without foundation. Change, nevertheless, will not come about merely because it would be beneficial; it requires work. Research on acceptance is now of comparable importance to research on methods for increasing the supply of food.

FURTHER READING

Bell, G. D. H., Moss, C. J. and Williams, O. G. (eds.) (1973), 'A Discussion on Agricultural Productivity in the 1980s'. *Phil. Trans. Roy. Soc.* (*B*), **267** 1.

Björkman, O. and Berry, J. (1973), 'High-efficiency Photosynthesis', *Scientific American* 229(4), **80**. Gives a clear account of C3 and C4 plants.

Ceres: a periodical produced by FAO (Rome) containing general articles on many aspects of agriculture and food production.

FAO, *The State of Food and Agriculture.* An annual volume containing general articles and condensed statistics.

Jones, J. G. W. (ed.) (1973), *The Biological Efficiency of Protein Production*, Cambridge University Press.

Milthorpe, F. L. and Moorby, J. (1974), *An Introduction to Crop Physiology*, Cambridge University Press.

Pirie, N. W. (1972), 'The Direction of Beneficial Nutritional Change', *Ecology of Food and Nutrition* **1**, 279.

Pirie, N. W. (ed.) (1971), *Leaf Protein: its Agronomy, Preparation, Quality and Use*, Blackwell, Oxford.

Pirie, N. W. (ed.) (1975), *Food Protein Resources*, Cambridge University Press.

Rechcigl, M. (ed.) (1973), *Man, Food and Nutrition*, Chemical Rubber Co. Press, Cleveland, Ohio, USA.

CHAPTER FOUR

AGRICULTURAL CHEMICALS AND THE ENVIRONMENT

G. S. HARTLEY

Introduction

The nature of agricultural chemicals

Among professional agricultural advisers and researchers, and within the chemical industry, the term *agricultural chemicals* is usually held to include all pesticides, fungicides and herbicides—most of which are industrial chemical products. Also included are various products which, again, have potent physiological action but are used for modifying plant growth (dwarfing of vegetative growth, setting of fruit, initiation of root growth in cuttings, etc.). Products for veterinary medication are not normally included in the group, except where their function is to reduce attack by external parasites; but, since some antibiotic products used in massive doses in intensive pig and poultry farming have been shown to have an effect on the human consumer of the meat, they have been included within the scope of this chapter. The products of heavy chemical industry or of direct mining of deposits—used in agriculture in much greater quantity to supply major plant nutrients (nitrogen, phosphorus, potassium)—are not generally included, because the scale of their use dictates rather different methods of processing and transport. Largely because of their massive use and, in the case of nitrate, ease of leaching into drainage water, they can have more impact outside their target area: they, also, must therefore be considered in the present context. Moreover, fertilizers are used in a wide range of dosage; lime often in much higher dosage than the N, P, K sources; magnesia in lower dosage but in a dosage which is higher than that of most pesticides; 'trace' elements, where deficiency occurs, at dosages which may be even less than that of expensive organic pesticides. Some of the most spectacular improvements of hill pasture in parts of New Zealand have followed the application of as little as 100 g molybdate per hectare, along with a normal phosphate dressing. It is required by the nitrogen-fixing nodule bacteria which enable clover to grow in the absence of available nitrate. Should molybdate be considered as a fertilizer or as a cure for a deficiency 'disease'?

In the present context it is best to define *agricultural chemicals* to include all those products of chemical industry—whether wholly synthetic or extracted from vegetable or mineral sources—which are applied in modern agriculture for the increased yield or other improvement of crops, but which may have significant effects outside their intended place or time for action.

The environment

We must equally reconsider the scope of the word *environment*. The question is of particular significance in an agricultural context because, to a high proportion of urban dwellers in 'advanced' industrial countries, the environment is the open country, and the open country (the openness of which raises wide political questions) is mainly agricultural and has been modified, even moulded, by agricultural practices over many centuries. At the present time of concern about the nominal 'balance of nature', this point must be emphasized. The author has met no better definition of *agriculture* than that Sir Vincent Wigglesworth once gave: *the art of disturbing the balance of nature most safely to our own advantage*. Industry can claim with some reason that its disturbances are much more localized, or at least can be made so, with some effort and cost, and perhaps compulsion. For many of us, agriculture modifies almost our whole environment. It almost *is* our environment. This places a great responsibility on those who practise and service agriculture. It should, however, suggest to spokesmen of the urban dwellers and workers the necessity to inquire what they really mean by the environment and the need to be fair and understanding in their complaints about disturbance. The picture of the healthy countryside or balanced nature conjured up by the townsman will, on closer inquiry, usually turn out to be the agricultural scene of his childhood days or even of his grandfather's time; but it should be remembered that the sun-drenched hay has been garnered from what was once swamp, the rolling uplands have been shaved for years by sheep, and the wild boar and wolf have been hunted out of the ancient forests which, in turn, have been reduced in various stages. All of this has happened under the relentless pressure of human population.

The so-called 'balance of nature'

Having said something about *nature*, let us consider further the subject of *balance*. There is a basic fallacy in the use of this word to describe the countryside of the British townsman's nostalgia. At its best—and this applies throughout much of Europe—there is a sense of the village or town having come to terms with its surroundings more amicably and in a more stable way than in the newer countries of the world or in more tropical climates. We may welcome all attempts to preserve this blissful marriage, but let us not call it a 'balance of nature'; it was a balance imposed by man at his best, building with care and love, and cultivating with tradition and experience—but imposed on what would otherwise be a wild scramble of competing species. If man departed, nature would take over his stable farmland in a dynamic succession of different flora and fauna, building up in some fifty years to a stable cover of beech or oak forest—stable until the disruption of some catastrophic flood or fire or earthquake. The natural balanced climax forest would be of little use in the British Isles to feed 50 million people. Agriculture does not disturb this balance; it destroys and replaces it. It is on the scrambling changes of the ecological succession that agriculture of a conservative type has achieved some balance between disciplined cultivation and nature's abundant capacity for change. There is a strong case for making changes in agricultural practice slowly, as all the consequences of change can never be foreseen. Conservatism is necessary.

Conservation

Conservation is a different matter. Change is inevitable. We may be able to make it desirably slow, but to hold any area in its previous stage needs some purely arbitrary choice. For most of us the choice, irrationally, is the time of our youth. Nowhere is the arbitrariness of conservation so clearly evident as in the writer's present environment in New Zealand. This country is, geologically, very young but, as a separate land mass evolving its own flora and fauna, very ancient. Its flora was largely unique and evolved in the absence of herbivorous mammals. Its native fauna included many species of flightless bird, mostly now extinct. Some of the extinctions—notably of the giant moa—were brought about directly by man, the hunter; more have been brought about by his introduction of other predatory species. Cats and weasles, introduced to control the introduced rabbit, found much easier game. Pigs introduced for food, goats for milk, opossums for fur, and deer for sport, have crowded out other animal species and are making drastic changes in the still large areas of remaining native forest. Introduced plant species have become serious weeds, the most spectacular being gorse and broom. While mostly these have followed cultivation, some, e.g. himalayan honeysuckle (*Leycesteria formosa*) and a mediterranean heath (*Erica lusitanica*) are competing for places suitable for their habits in the native scrubland.

Mount Egmont National Park must be one of the finest living plant museums in the world, including within its almost circular block of 130 square miles the complete range from lowland rain forest to alpine herb field. Its wardens have, by great vigilance and not without acrimonious opposition, kept out the deer. Feral goats are an even worse menace to trees, but the yearly shooting of 5000 is barely holding this menace in check. Elsewhere there are fears among other interests that some species of deer are being hunted out, and efforts are being made to preserve them. A current attempt to eliminate an introduced wallaby from an island in the Hauraki Gulf is meeting opposition from sentimental residents. One man's pet is another man's vandal. The outcome is not clear but, unless the conservationist lives in a country with a culture ancient enough to have forgotten the history of its natural history, his first problem is to decide what he should conserve.

Agricultural chemicals and other practices

Agricultural chemicals of all kinds have had a major influence on agricultural practice over the past fifty years. There has, however, been much interaction with other factors of change—increased mechanization, electric fencing, artificial insemination, decreased labour force, new techniques of plant breeding. Partly as a result of consequent increased specialization and more rapid transport, there has developed a more complex system of controls of quality. It would be very difficult to isolate and evaluate the part played by chemicals alone but, in so far as the changes are within agricultural practice, their discussion does not belong here. This is not a chapter on land use. If the townsman wants a different countryside for his children to play in or to serve as a backcloth for his caravan holiday; if he thinks that agriculture, or forestry, has become too dominant over amenity interests and, despite this, wants cheaper food, he is raising important political (in a non-party sense) issues, but they are general land-use issues not appropriate to this chapter. It is only when the use (or abuse) of chemicals brings them outside their time and place within the practice of agriculture that they can properly be considered to have their own environmental impact.

It is necessary to try to make this distinction because of the vocal and emotional prejudice which many people have against chemicals wherever used. They would not attempt to argue with the farmer about his selection of disc harrows rather than the plough, caterpillar versus wheeled tractors, hay versus silage, etc., but any use of chemicals is by many considered evil or at best suspect. This widespread prejudice is not logical, but it is a factor of importance which has had its effect on industrial policy and government legislation. Many may feel that it has been allowed to have too much effect, and that legislation has been more influenced by the emotions of the ignorant than the logic of the expert. In so far as there is an element of reason in adverse popular reaction, it must arise from a greater scepticism of expert knowledge when the expertise is in the use of chemicals rather than in that of machines or plants or animals. Particularly there is scepticism about the possibility of restricting the action of chemicals to a particular time or place.

Difficult decisions

If agricultural chemicals confined to proper use on agricultural land are not considered as entering the environment, should we consider not only the possibility of their escape but also the possible entry of persons on to chemically treated land? In most countries, the law of trespass does not make it an offence for the general public to enter fenced or cultivated land, but only holds the trespasser liable for proven damage. Although it is not an offence to enter, however, the public has no *right* of access. The farmer would not normally be held responsible if injury occurred to a trespasser from an animal or machine used on land enclosed in a fence intended to prevent the contact which caused the accident. The law relating to damage to a trespasser by a chemical used on private land may never have been made clear in the courts. The possibility of a trespassing picnic party picking up on food or clothing a significant dose from some newly sprayed field is, however, remote.

It is mainly the automatic dispersal of chemicals—dissolved in moving water, vaporized in moving air, carried by moving animals—which can create an environmental problem in the sense in which we propose to use the word *environment*, and it is scepticism about the possibility of controlling this movement which is one basis of the emotional anti-chemical reaction. There is another. Obviously some dispersal is inevitable. The expert considers that, from permitted practices, the escape is insignificant. The objector is again sceptical about the non-effect of officially tolerated levels of residues. Has he valid reason for his doubts? There seems, in modern political jargon, to be a 'credibility gap'. Is it due to ignorance of chemistry, bad public relations, bad evidence or, as the most extreme critics of the industry claim, irresponsible pursuit of profit?

Agricultural chemicals and carelessness

It is not proposed to deal with illness or fatality due to gross carelessness of individuals, major spillage-type accidents or the use of pesticides for suicide. Firearms and automobiles have caused far more casualties in these categories than pesticides, but such casualties are not normally considered as indicative of an impact on the environment.

Most people would accept, in general terms, that the chemical manufacturer should no more be blamed for casualties arising from accident, gross negligence, suicide or murder involving his product than should the manufacturer of ladders be blamed for casualties (far more frequent) arising from ladders being insecurely placed. Nevertheless, protests about an unfortunate number of fatalities arising from accidental consumption of liquid concentrate of the herbicide paraquat have been intemperately levelled at the inventors and producer rather than the domestic practices that made the mistakes possible. This substance is probably the most useful of herbicides on a small scale in the private garden. The convenient liquid concentrate, dark brown in colour, is in most countries only available in agricultural quantities. The 'helpful', or acquisitive, local jobber buys 5 gallons and splits it up among his 'friends' or small customers and uses, for the purpose, empty bottles. The efforts of protestors and reformers should be directed against these irresponsible practices, not against the producers of a substance which, properly used, has very valuable effects and is perfectly safe.

It is worth while noting that, among hospital admittances caused by accidental or wilful poisoning, the agent in the vast majority of cases was not a pesticide but a medicine (carelessly stored and permitting access by children), a disinfectant, detergent, solvent or chemical cleaning agent. In New Zealand in 1967, and the proportions are fairly representative of developed countries, poisoning of all kinds expressed in thousands, accounted for 4 out of 234 total hospital admissions and 2 out of 12 admissions following accidents in the home. These figures give a useful, if horrifying, indication of where action is most needed if we really wish to reduce accidental poisoning.

Chemistry and man

A plea for more knowledge of chemistry

Chemistry is still far too exclusively a subject for specialist chemists. The most important single step which could help to a more rational view of the problems under consideration would be to teach chemistry more universally and in a much more general form in schools. The basic message is that we are all chemists, if unconsciously so, breaking and making links between atoms in order to change one substance into another with slightly or wholly different physical properties. Our unconscious chemistry is, even today, more clever than the mentally-directed operations that go on in our laboratories and factories. More appreciation of the importance and versatility of essential vital chemistry would help to a more balanced appreciation of the uses and hazards of poisons, which are substances which can, in sufficient amount, seriously upset some vital chemical process. They are ever present from natural sources. This is not a subject from which we can escape. We can be logical or prejudiced, but not indifferent.

Living organisms, including man, can carry out the chemical reactions essential to life under a wide variety of conditions, while consuming a wide variety of food. Green plants can start from the most primitive chemicals because of their ability to use solar energy to reduce the terminal compound of carbon in animals—the dioxide—back to more complex and directly useful substances; they operate on other simple 'mineral' sources, too, and are the great primary synthetic chemists.

Animal chemistry is, by comparison, a second-hand sort of job—mostly using ready-made fuel to shuffle round the constituents of ready-made proteins and fats. The protagonists of 'organic farming' (who have many worthy motives for which we can have much sympathy) are, unintentionally, rather insulting to the clever green phoenix which is always rising from the ashes of the world. It alone does not need the products of organic chemistry; it makes them.

Poisoning is essentially quantitative

The needs of an organism are more or less fixed; the supply is more variable. Balanced growth is impossible without selection involving overconsumption and rejection. There are, of course, limits to the ability to select, and some species are more adaptable than others. Where excess of some supplied chemical is harmful to growth, or even fatal, we speak of it as being poisonous. Poisoning is essentially quantitative. All of us, Olympic medallists, Nobel prizewinners, ordinary folk, and the mentally or physically deficient, have consumed many times the amount of several chemicals which, if administered as a single dose, would have been fatal. We could not have survived had we not been able to deal with these substances, some of which, in small amount, are essential.

Natural and unnatural poisons

As far as the danger from industrial and, particularly, agricultural contamination of the environment is concerned, the reply frequently made to confession of faith in the versatility of the animal's internal regulating chemistry is that it may be justified for natural but not for synthetic poisons. The organism has evolved in contact with natural poisons over countless generations and has been able to breed adapted strains. The danger of modern pesticides is that they are new compounds unknown in nature. Both parts of this reply need qualification.

When the toxic properties of monofluoroacetic acid were first discovered about 1940 in Britain, this compound, structurally very simple, but requiring high temperature and pressure for synthesis and, when made, very stable, was hailed as a wholly new toxic weapon (by those concerned at the time with the possibilities of chemical warfare). It was difficult to detect either before ingestion or after death. It was a most unlikely compound for nature to produce. It so happened that, at the same time but without exchange of information, two South African biochemists were trying to identify the agent responsible for many fatalities in cattle having access to seasonal lush growth of a wild plant on the ranges. They were almost defeated, not by the complexity of the compound but its simplicity. For a long time it no more occurred to them than to the English chemists that nature could produce anything so ridiculous, but she did: fluoroacetate was the answer. He would be a bold man who would assert with confidence of any laboratory artefact that it does not occur in nature.

More important, however, than this real doubt is the certainty that the biochemical process with which the compound interferes is a natural process, probably subject to some natural interference, and certainly having its natural method of defence. It is doubtful whether any natural organo-phosphorus compounds have the type of biochemical action characteristic of the well-known insecticides of this group, but the biochemical disturbance is one which some naturally occurring poisons of different structure do create. The necessary defences are repair of the enzyme attacked, i.e. removal of the intruding chemical group, or replacement of the enzyme. The latter is very general. Enzymic chemistry is vital to the organism, and it could not survive without a mechanism for replacing enzyme molecules which have become non-functional. The rate of replacement is mainly a basic property of the organism and the enzyme system, and is largely independent of how the enzyme concerned has lost its function. Rate of repair (the reversal of inactivation) is very dependent on the kind of damage and therefore on the damaging agent. Even an unnatural poison is damaging a natural process and in so doing is calling into action a natural process of replacement, and perhaps also of repair.

The second qualification to the reply of the 'naturist' against synthetic poisons—that life has adapted itself to natural ones—concerns the completeness of this adaptation. It is indeed far from complete. More often than not, animals have had to learn to avoid a poison, rather than become able to cope with excess of it biochemically. The rabbit closely grazes the harmless herbage around the upstanding *Belladonna* plant; the American

Indian boils his manioc root and squeezes out the juice; the insectivorous bird avoids the brilliant colours of the toxic butterfly (and, incidentally, some harmless mimics too). Most cunning of all are some small sea-slugs which prey upon venomous jelly fish, avoid breaking the microscopic stinging cysts of the latter and utilize them for their own defence![1]

Two objective accounts are helpful in this connection. The following is quoted from the first.[2] 'The incidence has varied from 1 to 25% (of the exposed population).... The anomaly is restricted to the head and varies from a complete cyclops (fusion of the eyes and sockets) to a slightly deformed upper jaw.... The monsters are usually born alive, but severe cases die immediately from suffocation due to incomplete development of the naso-laryngeal passage.' The affected species is the sheep, the causative agent a Californian range plant, *Veratrum californicum*, a close relative of a plant from which an insecticide was at one time extracted and considered to need no residue clearance as it was a natural product!

Two further quotations[3] refer to the human species.

'In a single district there are as many as 25,000 cases out of a population of 634,000.'

'In most cases the onset of disease is sudden. . . . In the milder cases there is bending of the knees and difficulty in running. More advanced cases walk on the toes and require a stick. . . . This leads to the two-stick stage and finally the patients (mostly men between 20–29) are reduced to crawling.'

The causative agent is *Lathyrus sativus*, a heavy-yielding drought-resistant pulse (relative of the garden sweet pea) eaten in some poorer districts of India.

It is customary to record doses of poisons ingested by (or otherwise administered to) experimental animals in terms of mg of poison per kg of body weight, larger animals for obvious reasons needing more to initiate equivalent response. The lower the figure, the more toxic the compound. The most usual standard of response in the case of acute toxicity is 50% deaths in the population studied. Few permitted insecticides have, on this scale, an 'LD_{50}' less than 10. A very few are as low as 1 or 2. Several synthetic poisons are known with values well below 1 but, until recently, the record lows were scored by two natural poisons—the toxic product from meat invaded by the *Botulinus* organism and 'aflatoxin', produced by the mould, *Aspergillus flavus*, growing on ground nuts. Now the synthetic impurity associated with 2,4,5-trichlorophenol, namely symmetrical tetrachlorodibenzodioxin holds the record, but only in the guinea pig.

Such figures are often allowed to influence decisions in a more absolute way than they should. The risk depends on concentration and usage, or occurrence, as well as inherent toxicity of the pure compound. A pesticide of $LD_{50} = 200$ but used at a rate of 10 kg/ha is, on a first calculation, more of an environmental risk than one of $LD_{50} = 20$ but used at only 20 g/ha. The intrinsic figure for neither an impurity nor a natural product should be used for alarmist propaganda without stating the proportion existing in the industrial concentrate or in the growing plant. Many other factors, too, are involved in the environment risk, but this is no excuse for starting from unrepresentative quantities.

Acute and chronic effects

The measure of toxicity just referred to is usual in the examination of acute effects following a single dose. To make the dosing more precise, although less realistic, data are often quoted for intravenous injection. The method of administration should, of course, be specified. In practical assessment of toxicity to many pests we have to use much less direct measures, such as amount sprayed per unit field area.

Acute toxicity is of great interest to the biochemist, the manufacturer, and the user in the field. It is not the most important aspect for environmental effects—risks of what may happen outside the treated field either to the inhabitants of nearby houses, the drinker of drainage water (suitably processed) or the consumer of the crop. Only occasionally, and with no good reason, has anyone suggested that any of these could receive a single lethal dose, or anything near it. There is, however, a perfectly reasonable question about long-term effects of small intakes over a long period.

Long-term low-level toxicity, and acute response to a single dose, generally called 'chronic' and 'acute' toxicity respectively, both require fuller description in serious discussion. They are not necessarily closely related. It may not even be possible to express a quantitative relationship, because the single large dose may evoke quite different symptoms from the repeated much smaller dose—the effect of selenium on a deficient pregnant ewe is a good example. (Spectacular results in New Zealand followed the application of traces of selenium, a notoriously toxic element, to pastures. Without selenium, ewes in apparently good health fail to carry live lambs to full term.) Even where the symptoms are more or less the same, the ratio can vary widely. The relation of acute to chronic toxicity depends on the speed with which damage can be repaired, the length of time that one process in a chain can be disrupted without disturbance elsewhere, and the speed with which the toxicant can be degraded or excreted. Since reaction rates are involved, variation between different toxicants can be very great.

At one extreme we have the classic example of an acute non-chronic poison—cyanide. It has a moderately-low acute toxicity, and its lethal action is very rapid, but it is almost insignificant if administered slowly. Cyanide, in fact, is an inescapable constituent of normal diet. *Almost*, cyanide either kills you or does you no harm. It comes nearer than any other chemical to the 'hit or miss' simplicity of a rifle bullet.

The opposite extreme is provided by lead salts. The acute toxicity is moderate, but it matters little whether the dose is taken in over minutes or months. There is mobility within tissues, but both excretion and the effect on the brain cells is slow.

Acute and chronic biochemical mechanism may not run parallel with overt symptoms. A suddenly inflicted and perhaps irrecoverable damage may take a long time to exert its physiological effect, and this may itself take a very long time to repair. A case in point is the prolonged, perhaps permanent, paralysis which has followed acute intoxication with some organophosphorus compounds. This very serious symptom has never been demonstrated, except as a sequel to near-lethal acute symptoms. Although physiologically chronic, it is probably therefore of no importance in assessing environmental, as distinct from industrial, hazard where we shall always be concerned with low-level but perhaps prolonged intake.

The acute-chronic difference is not a simple one. In experimental work, particularly on short-lived species, such as most insects, it is necessary to be clear about how dosage rate is measured and to distinguish between the time periods involved: duration of exposure, time interval between exposure and appearance of symptoms, duration of symptoms.

Significance of persistence

Failure to realize beforehand the significance of stability and persistence in the case of the first very stable insecticide of the 'synthetic era'—namely DDT—has been more than any other factor responsible for current alarm over the environmental hazard. DDT, at the beginning of its history as an insecticide, was known to have a very low acute toxicity (i.e. a high LD_{50}, around 120 by mouth in the rat, corresponding to 8 g in one dose to an adult human) and to be devoid of any irritant action on contact, even with the eyes. It was easily made and very stable. These virtues at once gave it an important place during war-time in the control of insect-borne diseases, particularly malaria and typhus. This was a time when short-term urgencies dominated thinking, and DDT on a large scale did a very important job. Its war-time success (and the existence of stocks and manufacturing facilities) carried it into a euphoric period of peace-time use.

Some undesirable results of the dispersal of enormous quantities of DDT could have been anticipated. Others should at least have been considered possible, and usage therefore slowed down; but by whom? It might have been wiser for the stocks, produced primarily for war needs, to have been 'frozen' in the same way as stocks of mustard gas and other chemical weapons, but mustard gas was a poison, DDT was an insecticide.

It is customary to speak of unwanted and perhaps unanticipated effects of a pesticide as 'side' effects. Six major effects of widespread DDT usage may be listed and all but (1) and (3) are side effects. An advance on a narrow front has very long sides!

(1) An unprecedented reduction in human mortality from malaria in regions where this disease was a main cause of death.
(2) A resulting almost impossible population explosion because the natural control was eliminated before any social control has been established. Since population pressure is the first factor in pollution, this is probably the worst act of pollution attributable to DDT.
(3) Very material improvement in the yield of many crops.
(4) Development of strains in many insect species much less sensitive than the original.
(5) Replacement of one pest by another, where a parasite or predator has been more sensitive than the pest. Red spider in orchards is the classic case.
(6) Accumulation in the fatty tissues of many animals, particularly those, including man, high in a food chain. Raptorial birds have aroused most interest.

DDT was followed by other insecticides of what has come to be known as the chlorinated hydrocarbon class, including aldrin and isodrin and their epoxides dieldrin and endrin and their relatives. These are chemically even more stable, and their very great persistence has caused their use to be severely restricted in most countries. There is now strong pressure to get DDT completely outlawed.

Benzene hexachloride (BHC: the active γ-isomer is called lindane in most countries) was not included in the last paragraph, although it is a chlorinated hydrocarbon and the next to come forward after DDT. It is much less persistent than the others, and it should not be included with them.

To the active opponents and temporizing politicians, the greatest crime of these insecticides followed from their extreme persistence—point (6) of the list. A high partition ratio in favour of fats from water contributes to the stability (again less with lindane than the others) because most reactions occur in water with other compounds active therein. A compound stored in fatty tissue is in a relatively inert environment. When fats are consumed by an animal, they are split up in the gut and remade or further metabolized after passage through its wall. If metabolic demand is high, the less reactive DDT (or other inert fat-soluble compound) accumulates to a yet higher concentration in the proportion of fat passed on for storage. This is the plausible mechanism of the famous accumulation in the food chain, but some serious students of environmental problems consider[(4)] that it is not clearly established that the occasional very high and damaging content of DDT in some raptorial bird carcases has been built up in this way, rather than by the bird feeding on an accidentally rich source—such as a sheep killed by an accidental overdosing of insecticide.

A great deal has been written about the dangers of excessively stable insecticides, and the reader is referred to two well-informed discussions.[5,6] Although all parties now accept that it is in principle undesirable for the fatty tissues of a man to carry as much as 10 ppm DDT, there is no evidence that this has done any harm. We can realize, with horror, that the compound is so stable that this is possible or, with relief, that it is so inert that only the analytical chemist can tell us it has happened. A more important environmental effect for all of us is that the very persistent compounds are difficult to exploit selectively, and that during their dispersal many insects are exposed to a wide range of dosage, a condition best suited to development of resistance.

There is still political (in a non-party sense) argument between restriction of use and total ban. The protagonists of banning consider that restriction is too difficult to enforce, because the control of final dispersal of a pesticide is in too many hands in too many remote places. They are given not-always-welcome support by the emotional protestors for whom slogans longer than 'DDT out' or 'ban 2,4,5-T' are too long to shout. The restrictors argue that the evils have followed unintelligent use, and that the persistent insecticides are weapons which could safely be used in more intelligent ways. A lot has already, if painfully, been learnt about them. It might be better to put more effort into their better use than into the search for new compounds in which known bad properties are less but which may have other properties we do not at present suspect. Against this argument, the banners can claim that we are now in a better position to anticipate trouble than when DDT came in as a war weapon with little thought of 'side' effects. A large proportion of the greatly increased global research effort is devoted to the study of residues, testing for development of resistant strains, effect on predators, and low-level long-term toxicity, including carcinogenic and teratogenic potential.

Persistent and transient compounds

Chemical industry has fully accepted that there is no future for compounds as persistent as DDT. No new compound of such stability would now even pass from the research laboratory to the next, more expensive, stage of trial development. A compound can, however, be so transient that it would be quite useless in agriculture. Only when pesticides are applied directly to the target, e.g. insects exposed at the time of spraying, can they be useful without some persistence. Most herbicides must be translocated if they are to do more than kill off exposed leaves; many are applied to soil to kill seeds as they germinate and must persist for a few weeks at least. Systemic action of insecticides in plants, the most effective way of attacking hidden insects, needs some persistence because transport is slow. Residual effect is required to kill larvae emerging from eggs.

The problem therefore is to have compounds which are persistent enough to exert their useful effect without being so persistent that they create environmental problems. This is made much more difficult by their being used under necessarily variable conditions, particularly of temperature and moisture availability. If the ideal effective life of the compound for one intended purpose is, say, three weeks, it could easily be several times this if the weather becomes exceptionally cool and dry, or its effectiveness will be lost too early under exceptionally warm and moist conditions. A fairly wide practical tolerance between the ineffective and hazardous periods is required.

Most toxic compounds are of limited life under field conditions, if only because they undergo chemical reaction in causing the biochemical disturbance responsible for their toxicity—their molecules, like the bee, commit suicide in the act of stinging. This is true of the organo-phosphorus poisons, and most compounds where the mechanism is known. Even fluoroacetate, a very stable substance in an inanimate environment, undergoes reaction to the much less stable fluorocitrate in blocking transiently the vital tricarboxylic acid cycle. It is, in consequence, short-lived in animals and some plants, and is attacked rapidly by soil bacteria. The example is not unique. Tetraethylpyrophosphate (TEPP) has a maximum half-life in water (in which it is freely soluble) of about 8 hours at 20° but is highly toxic while it lasts. Octamethylpyrophosphoramide (schradan) is almost indefinitely stable under the same inanimate conditions. It is converted by oxidation in animal tissues to a compound at once much more rapidly destroyed by hydrolysis but much more toxic. The conversion takes place more slowly in plants but leads to inactivation in a few weeks. A great deal of selectivity of useful pesticides depends on this type of behaviour. They have different possibilities of reaction, by vital and inanimate mechanisms, some leading to a toxic effect and others to harmless inactivation. The relative speeds of these reactions (and of rates of access to the site of the toxic reaction) can make one compound relatively more toxic to mammals, another to insects.

Some compounds can exert toxic effect without decomposition. This is the troublesome peculiarity of the DDT, aldrin, etc., group of insecticides. The molecule appears to inhibit sodium ion transport across nerve sheaths by being adsorbed on to and blocking the carrier channels. The 'hormone type' herbicides probably also have an essentially obstructive rather than reactive mechanism, but they are subject to other chemical attack.

A great deal is known about the chemical breakdown of pesticides, and the general public should be more aware of this. It may have been rather innocently true in 1944 that DDT was used on a large scale without any knowledge of its chemical fate. The situation is greatly changed in 1974. An irregular journal[7] has produced over 50 volumes of expert papers in this field. In 1969 there appeared a multi-author text[8] on 'degradation' (chemical decomposition in water, soil and plants) of herbicides only. This ran to 350 pages of text and about 900 specialist references, nearly all since 1950. It has already needed revision.

The enormous increase of research effort which this illustrates has been stimulated by legislative requirements. The marketing of new products is in most countries prohibited until a great deal of experimental evidence is presented to, and approved by, the appropriate government department. Even in those countries where the examination is still nominally voluntary, it is effectively compulsory for two reasons. One is the significance of export trade, e.g. the use of DDT on pastures has been banned for several years in New Zealand, although it is the best control of grass-grub, because meat, cheese, etc., would not be acceptable in the United States and many other potential importing countries if they contained significant DDT. The other virtual compulsion is the impact of world public opinion, all the more important because so often ill-informed and irrational. The manufacturer prefers to have government approval, even where it is not legally required, since it gives him the strength of a second, and impartial opinion.

Yet another interest now favours energetic research by the manufacturers into safety and the environmental fate of compounds. Safer and less contaminating compounds are going to be commercially more successful. At the biochemical level, safety, selectivity and environmental fate are all closely related. The investigation of one greatly helps that of another.

Presence at the wrong time

Contamination outside the intended time of action can occur as a result of overdosing, premature harvesting, gross carelessness or, to take a probably rare example, theft. None of this is under the control of the consumer. It is here that there is an important difference in principle from the other much-used physiologically active chemicals—medicinal drugs. If you overdose yourself with a drug, it is your fault, or possibly your intention. If you take a drug curatively under medical advice, you have some choice in balancing a risk against the inconvenience or danger of continued ill-health. If you buy vegetables or meat, you have no personal control over their prior treatment.

You may have the opportunity of avoidance, in the case especially of vegetables, by growing your own. In an urban and industrial society this has no general significance, and some domestic growers use insecticides more carelessly and liberally than any commercial grower could afford to do, either for his pocket or his reputation. Others prefer a healthy population of pests. In a general way then has the consumer a choice? If the housewife were not so insistent (or were not assumed by the suppliers to be so insistent) on unblemished leaves and fruits, might she have less risk of insecticide? De Bach[9] makes the point well in his final editorial paragraphs of a major textbook on biological control. Are we, in our anxiety to keep pesticides out of the environment, prepared to let a little more environment into the larder?

This conflict is becoming increasingly important with the increased market in prepacked and frozen vegetables. The purchaser of loose vegetables from the greengrocer is prepared to reject some blemished fraction, if only the outer leaves of lettuce, cabbage, etc., left on as moist wrapping materials. The contents of the prepared package are expected to be ready to cook and serve. Dead insects are at once a cause for rejection and complaint. The grower for the package trade has therefore to meet a more exacting standard. The standard *is* being met as is a very safe standard for insecticide residues; but the dual standard has meant increased cost, and the cost is inevitably rising as resistance to cheaper insecticides necessitates the development of more expensive ones.

In a sophisticated urban society we have to take a great deal on trust with regard to safety of purchased products. With present regulations, enforced by sampling and analysis, it is doubtful if we are any more at risk from the greengrocer than from the furniture, toy, textile and dry-food stores. Most of the goods we buy in these places have received at some stage some chemical treatment for protection or to facilitate processing. There is here no attempt to excuse by accusation, but only to make the point that most of us have been satisfied for a long time with the machinery to protect us from the harmful processing of these other goods. It is not logical to assume that contamination of raw foods is likely to be less well controlled.

Presence in the wrong place

Presence in the wrong place is the particular environmental hazard of agricultural chemicals. They must be spread over large areas at a dosage level highly significant for some form of life at the time of application, whether this significance is improved growth or reduction of unwanted population. They can spread from the target area by drift of particles during a spraying or dusting operation. They can evaporate into the moving air and percolate into rivers.

Aerial drift

Aerial drift is the most evident of these risks. Its effect is most evident with herbicides. Little problem will arise on neighbouring fields from a minor trespass with fungicides or insecticides, except in unusual situations, such as Southern Asiatic Russia, where mulberry trees serve the dual purpose of providing raw material for the silk industry and wind-break for cotton, which is very vulnerable to insect damage. In mixed cropping areas, however, it is common for a crop on one field to be more sensitive to herbicide than the weeds in a nearby cereal field. Tomatoes and vines adjacent to cereals are particularly at risk from 'hormone' herbicides. The problem is aggravated by the hormones producing significant symptoms at a dose far below that causing permanent damage and, by these symptoms, in the tomato plant, being not unlike those due to exuberant growth with excessive nitrogenous fertilizer. Compensation for drift damage is an important and specialized branch of agricultural insurance; false claims in a glut season are not unknown and are difficult to deal with.

This type of problem is, however, a local one. It has given rise to acrimonious disputes, but generally responsibility is clear and the spray operator has the important reason to take precautions that he can incur severe financial loss. It is perhaps no more a general environmental problem than age-old trespassing of the grazier's animals on to the arable farmer's land

Despite its importance, basic study of the mechanism of drift is very lacking. There are several reasons for this, and they are particularly worthy of comment in the present context. (1) It is a subject where some rules seem obvious and therefore are not questioned. (2) Except for the sale of some formulation additives claimed to reduce the risk (and probably rightly) and of special mechanical techniques, advances in neither of which directions are easily protected by patents, no one can make money out of it. (3) Its prevention is not a clear responsibility of any government department. It is, in political jargon, a buck which is always being passed.

It is obvious, and true, that aerial spraying creates more drift hazard than spraying by ground machine. It is obvious, and true, that spraying at ultra-low volume, because it needs smaller droplets to obtain adequate coverage, creates more environmental hazard than high-volume spraying. This trend tends to magnify the first because, owing to the high cost of lift, ultra-low-volume spraying is most advantageous from the air. By ultra-low volume (ULV) is meant spraying of the concentrate itself, with only such minimal addition as is necessary to keep it liquid. The rate may be as low as 1 litre/ha.

As far as localized damage is concerned, the obvious case against fine-droplet spraying is not necessarily valid. Nor is the obvious advice to spray only when there is very little wind and, preferably, for this reason and because the air movement is usually not turbulent, in late evening or early morning. The arguments are based on rate of fall of droplets (in still air): below 100 μm diameter this rate is nearly proportional to the square of the diameter. It is often stated that, for this reason, a 50-μm droplet can drift four times as far under the same conditions as a 100-μm droplet. The logic is grossly over-simplified. In turbulent air, drops in this size range are carried up and down in irregular paths with little difference. A higher proportion of each will drift a much longer distance than simple consideration of horizontal carry and vertical fall would indicate. The difference in rate of fall only becomes operative when any turbulent parcel of air gets trapped within the canopy. *En route* there is more chance of the larger droplets being impacted on to surfaces and staying there; this behaviour depends very much on whether they are solid or liquid. While this is happening, the droplets are evaporating at a rate proportional to radius.

A more probable picture than the orthodox one is of droplets starting below about 100 μm being rapidly reduced in size by evaporation in ordinarily dry air. They may become solid; but this depends on the formulation. In moderate wind, these micro particles will be so far dispersed during inefficient impaction that they constitute very little next-door risk, although they do contribute to generalized pollution. In gently-moving air under inversion conditions there will be very little evaporation; droplets starting at the same size will drift a short distance and be efficiently collected. This picture is confused by the operation of opposing factors. There is a strong case for much more objective experimentation than the subject has received.

A good example of the low intellectual level of drift-thought is that advantage has been claimed for making the spray heavy. Rate of fall is proportional to density which, at great expense and with other attendant disadvantages, could be increased by a factor of 1·5; also to the square of diameter, so that an increase of 22% in diameter would be as effective. Even more remarkable was a transient commercial success of foam formulation to reduce drift, associated with a statement that foam was denser than spray! Foam application may have some local advantages if one can afford to use sufficient liquid to hold the foam down. Chief of these is that the deposit is more visible, an important advantage in spot spraying of perennial weeds.

A pesticide may be dissipated by the air in vapour rather than particulate form, a subject which will be dealt with in a later section.

Flowing water

A pesticide may be carried to the wrong place by flowing water. Indeed, if it gets into flowing water, it is in an important sense in the wrong place already. There has been massive and prolonged pollution of many rivers, and occasional serious incidents in others, when fish life has been destroyed and the major source of toxic pollution has been chemical factories and chemical-using factories. Often the trouble has arisen from discharge of materials not, on ordinary standards, toxic. They act by providing so much oxygen-demanding food for bacteria that the river becomes untenable by less-adaptable oxygen-dependent creatures. For obvious reasons, good food is, in respect of BOD (biological oxygen demand), a worse offender than many poisons. Uncontrolled 'swillage' from milk-processing factories and abbatoirs are the worst contaminants.

Incidents have arisen from spillage of toxic chemicals from riverside factories and even, in the case of small flows, from discarded unwashed drums or contents of sheep dip tanks. We have all seen what *Homo vulgaris* sees fit to dump in the village stream. These are incidents rather than the inevitable byproduct of chemical usage; their cure is in education and policing of control measures. Rarely have rivers been contaminated by pesticides as a result of their normal use, but there has been direct conflict between the salmon fishing interests in British Columbia and the requirement of the forest service for aerial spraying of DDT for spruce budworm control. There have been other conflicts between different users. A rather general one concerns the control of obstructing aquatic vegetation in navigable waterways. These may also be used for irrigation, and weed control is particularly wanted in irrigation ditches themselves. The herbicide must kill the obstructive vegetation, worst of which is the tropical water-hyacinth, without killing either rice or fish. These are essentially local matters between conflicting interests rather than one of general pollution.

Pesticides may occasionally gain access to rivers after their normal use on the land. Most pesticides, however, are strongly adsorbed on the enormous surface of organic and clay particles of soils, and their entry into drainage water is negligible. The exceptions are compounds which are rather simple strongly-acidic molecules, since the soil solids have little affinity for simple anions except for phosphate, which forms metal salts of very low solubility. Trichloroacetate, dichloropropionate and 2,3,6-trichlorobenzoate (and the related picloram) are the compounds which have occasionally been carried with undesirable effect into drainage water.

The two latter can produce deformity in sensitive plants at a very low level.

Surface run-off with associated soil erosion has produced a maldistribution of applied herbicides, in that the desired effect may not be obtained on the slopes but the persistent effect of the increased dosage may show its effects in the valleys. General pollution by this method is unimportant, however, since the water volumes ensure great dilution.

Important and undesirable transfer of the herbicide TBA has followed from the practice in Britain of growing tomatoes under glass, using straw bales as rooting medium. If these bales have come from a wheat field treated in early growth with TBA, poor and distorted growth has followed. Initially a wheat field may have received only 100–200 g/ha, but wheat is very resistant. A significant residue is left in the straw. The tomato, extremely sensitive to this herbicide, is growing on straw which occupied about 100 times its present area. This illustrates a particular 'wrong place' mechanism which can be avoided once the danger is known.

The most significant agricultural chemical percolating from normal use on the land into drainage water is nitrate. This can certainly contribute to 'eutrophication', i.e. excessive vegetative, particularly algal, growth. It must not, however, be assumed that this results only from the application of 'artificial' fertilizer. The nitrogen content of organic manures has to be 'mineralized' by bacterial action, i.e. oxidized to simple nitrate ion, before it is available for green-plant nutrition. Some wastage of nitrate, because it is not held up on soil solids, is inevitable. Land in good heart, growing heavy crops, is bound to release some nitrate to drainage. We must not expect a river fed only from agricultural land to be a sparkling undernourished trout stream. Mineral nitrate fertilizer can certainly increase the concentration in drainage water, and what escapes this way represents economic loss; the loss is greater when a massive application is made once or twice in the season, but frequent application is impracticably costly from the air or mechanically damaging to the crop when ground machines are used. Slow release formulations could be advantageous, but are at present very costly. Much research effort, however, is being put into this project. Two types of product are being developed: (1) insoluble high-nitrogen polymers, e.g. urea-formaldehyde condensation products, which require bacterial attack to be 'mineralized'; and (2) soluble salts coated with an envelope which is only slightly porous.

Contribution by drainage water of potassium or phosphate is negligible. If land is so rich in available phosphate as to contribute materially, it would be economically wasteful to apply more. Industrial and domestic sources of phosphate are much more important. It is by no means clearly established[10] that phosphate in treated sewage, mainly arising from polyphosphate in domestic detergents, is the most important factor in eutrophication. The search for alternatives, or even research as to whether their function is really necessary, should be subject to pressure for phosphate economy. Forseeable reserves of concentrated phosphate deposits are limited. Phosphate is essential for agriculture; it is much less necessary in washing powders.

Eutrophication has become an increasing problem in lowland rivers, and influx of chemicals, even influx of sewage, is by no means entirely to blame. Two factors of land-use, or water-use, policy are important. With increasing urbanization and population, water demand has increased, and cleaner mountain sources are no longer adequate: over-abundant growth is more a problem when water must be abstracted from lowland rivers. Secondly, increased utilization of lowland rivers, particularly when power also may be obtained, requires construction of dams and therefore a longer residence time of water in the lakes created. This encourages growth, restricts downstream losses, and at the same time makes the growth more undesirable for immediate interests, obstructing traffic, fouling machinery, reducing the life of filters. Encouragement of controlled vegetation of a type which could be dredged out for useful processing would seem a better answer than expensive attempts to kill wild vegetation by herbicides or macerate it to let it pass through turbines.

Long-range pollution

Most pesticides are significantly volatile, even those not used deliberately as fumigants. The fact that, in agricultural use, they are spread over a large area exposed to sun and wind necessitates a revision of the chemists' laboratory standards of volatility. 1 kg/ha (about 0·85 lb/acre), a quite high dosage of many modern pesticides, would, if spread uniformly over a smooth plate-glass field, make a layer only 0·1 μm thick. At 20°C it can be calculated that dichobenil, a soil-applied herbicide, would evaporate completely in about 8 hours, parathion in about 3 days, lindane in about 3 weeks, and DDT in about 9 months.

The field is not a glass plate and the spray is not uniformly spread. From isolated spray-residues evaporation will occur at comparable rates, but much of what is evaporated is collected again on other surfaces of the crop and soil. On leaf surfaces it can dissolve in the outer cutin layer and contained oils and penetrate into cellular tissue. In both this tissue and in soil, it can dissolve in the contained water and be adsorbed on the extensive internal surfaces. In these situations it is attacked, at very different rates according to its chemical structure, by vital and inanimate chemical processes. The amount eventually evaporating unchanged is thus dependent, not only on the volatility of the pure substance, but on various properties. It can be quite high, though always slower than from the smooth glass plate.

The most important of the other properties is solubility in water, because this is the most abundant solvent; generally speaking, the more water-soluble compounds are also more quickly reacted. Oil-soluble compounds are held up in leaf cuticle components but are more likely to escape without reaction.

Upper-atmosphere chemistry

What happens to the pesticide evaporating into the air above a single sprayed crop? Eddy diffusion will carry it upwards and downwards as it moves downwind. Since the surface of other crops and soil will provide a sink to adsorb and dissolve much of what reaches it, downwards diffusion will predominate. Upward, however, eddies become larger and more violent, and the induced diffusion process more rapid until, at a variable height of a few kilometres, the adiabatic temperature lapse gives place to a permanent inversion where air flow is more smoothly laminated. Molecular diffusion is increasing as the air becomes thinner. Above about 50 km, only 10^{-4} of the total weight of atmosphere is found, but about $\frac{1}{500}$ of the oxygen is present as ozone, kept in balance by creative and destructive reactions activated by intense short-wave-length ultraviolet light. The primary ozone-producing mechanism is photochemical disruption of oxygen molecules, and capture of the resulting atomic oxygen, even more reactive than ozone, by normal oxygen. The ozone layer is essential to the survival of living organisms at ground level, because it filters out the very damaging far-ultraviolet rays. There is no doubt also that it breaks up any organic molecules which living or industrial chemistry sends up to it.

We have therefore a perfect sink for chemicals above 50 km, the rate of access to which is difficult to calculate, and a variable less-efficient, but more easily accessible, sink at ground level. The ground-level sink can hold many substances transiently if they are not degraded by inanimate chemical reaction or by living chemistry in plants and in soil microbes. It is satisfactory that there is increasing expert interest in upper atmosphere chemistry,[11] greatly helped and stimulated by satellite technology. Some alarmist speculations follow objective study: they may be well founded, or their originators may consider that alarm is necessary to arouse interest. Depletion of ozone by nitric oxides liberated by supersonic high-altitude aircraft and by atomic chlorine derived from reaction of aerosol propellants have both been predicted. The upper atmosphere is where we must look for long-term pollution chemistry. It is a region where there is an ultimate clean-up of all organic misfit chemicals, but possible disastrous interference with the present chemical balance.

A few composition facts are not without interest to get a sense of perspective. The total mass of the earth's atmosphere is 5×10^{15} tons, more than a million tons per head of human population but about $\frac{1}{300}$ of the mass of oceanic water. After the major constituents: nitrogen, oxygen, argon, water and carbon dioxide, in that order of abundance, we come to minor permanent constituents, the most abundant of which is methane, closely followed by nitrous oxide, both mainly of vital origin, at 5 and 3×10^9 tons respectively. Ozone is next on the list at $1{\cdot}5 \times 10^9$ tons, but nearly all of it is above 50 km up. Sulphur dioxide is about 9×10^8 tons, and its origin fairly evenly divided between volcanoes, plants and industry. If DDT, the least volatile of the common insecticides (v.p. 2×10^{-7} mmHg at 20°C) could be dispersed into the atmosphere, about 10^7 tons (more than the total world production to date) could be held in vapour form.

Legislation and social responsibility

Since the wide and careless use of DDT in the late 1940s, governments of all countries have become aware of the dangers of uncontrolled use of agricultural chemicals. The chemical industry in most countries has been subject to control of what it may produce (it has long been subject to control of gaseous and liquid effluents) and has, for reasons already stated, willingly cooperated in some restrictions not legally enforced. The farmer and the agricultural contractor are much more aware of the ill-effects of some usages than they were thirty years ago. Their concern is more limited to disadvantages in the near future and in near places—drift damage incurring claims for compensation, residual herbicides depressing the yield of a following crop, poor control or upsurge of pests where predators have been eliminated. These near effects are greater than most remote effects; the general public, making only more remote contact, should realize that, as a good farmer can use chemicals to the clear and long-term advantage of his cropping, he is unlikely to depress health or fertility outside his farm. The fact that most risks are greater for someone else—manufacturer, contractor or grower—than for the general public, is a rather forgotten fact that should carry some assurance.

Of course there are exceptions. Global atmospheric and oceanic effects are outside the farmer's knowledge and control in a competitive world. It is where immediate competition forces action opposed to long-term interest that legislation is necessary. The main long-term interest which legislation attempts to protect is the health of the consuming public. Some insecticides have been banned and others severely restricted, because they are held to cause possibly significant contamination. A complete ban is much easier to enforce than control of use, because there are few manufacturers and many users. Legislation is necessary, because a grower using a contaminating chemical could offer an attractive product at a lower price. Legislation has also been introduced to protect particular interests, such as fisheries and wine-production, from operations of farmers or foresters who have no financial interest in them.

So far legislation has not attempted to enforce agricultural chemical practices which are contrary to immediate local interest but of long-term benefit other than safeguarding consumer health, yet it is here that positive evil can be established rather than remote possibility suspected. There are, for example, actions calculated to reduce the tendency to breed resistant strains which could not be effective without legal enforcement. Such actions are partial treatment of the crop, leaving whole fields or strips within them completely untreated to serve as reservoirs for normal pest insects and their predators, and alternation over a several-year cycle of the type of insecticide permitted. The first would not be popular with competitive farmers, the latter not with competitive chemical producers. It is difficult to see how these and other strategic devices could be implemented successfully while the pesticide industry remains internally competitive. It might be difficult to argue the case against a proposal to nationalize the pesticide industry, but such a proposal has not appeared in a socialist political programme. Nationalization would, in the author's opinion, slow down innovation, but this might be desirable. Perhaps the least desirable result would be that the whole activity, as is true at present of the safety regulations, would be too sensitive to ill-informed opinion.

Vietnam, 2,4,5-T, and the chicken oedema factor

The recent storm over 2,4,5-T provides an interesting example. High-level hearings are still pending before the Environmental Protection Agency in the United States, and it is uncertain what uses of this herbicide will be allowed to continue. There is space here only to mention some cross-currents. 2,4,5-T has been in use in agriculture and forestry for nearly as long as 2,4-D but on a smaller scale. Effect on the health of no one had been known or even suspected before its use by the US Army as a defoliant became one rallying point for general protest against the Vietnam War. In 1959, at first quite unconnected with 2,4,5-T, mysterious mass fatalities of chicks occurred in the United States and the incident was traced to a large batch of stearic acid incorporated into their feed. Compulsory biological testing for the 'chick oedema factor' in this raw material was instituted, and research to discover the cause initiated. This was successful in 1967 in tracing the toxicity to polychloro-dibenzo-dioxins. These came from high-temperature treatment of poly-chlorophenols with which the fatal batch of stearic acid (there have been no repetitions) must have become contaminated. Intensive investigations of toxicity followed, and it was realized that 2,4,5-trichlorophenol (from which the herbicide 2,4,5-T is made and itself produced by high-temperature alkaline treatment of s-tetrachlorobenzene) can contain s-tetrachlorodibenzodioxin, or TCDD as it has come to be called. One investigation established that TCDD (in dosage to mice much higher than could correspond to incidental dosage in man) caused hare-lip deformities in offspring. This was at once coupled with reports that the incidence of deformities in babies had increased in Vietnam since 2,4,5-T was used. Critical opinion of these reports leaves little doubt that they have no significance, but the damage was done, and there followed a number of quite baseless attributions of normal deformities elsewhere to possible contamination with 2,4,5-T. The TCDD content can easily be reduced to an insignificant level, once the desirability of this is known, and one large manufacturer had tightened its specification in this respect before the scare occurred.

Anyone doubting that popular emotional reaction can interfere with objective research and reasoning should read the research paper announcing the identification of the chick oedema factor and its abstract[12] in the most highly organized of information summaries. Although some of the authors of the paper are politically anti-2,4,5-T, their statement of findings and speculation on origin is fair and objective. They point out that polychlorophenols have many uses, and there are obvious possibilities of contamination. The abstract, however, attributes the occurrence in the fat concerned to '*animals feeding* on pesticide-treated *grass* and *fodder*' and none of the words italicized appears in the original article.

The chick incident should teach another lesson. There is little doubt that a single unknown incident led to the significant contamination of a single large batch of vegetable oil before it was subject to high-temperature hydrogenation. It is, I think, greatly to the credit of the chemical industry, dealing as it does with large numbers of chemically different but apparently similar liquids, that such accidents are rare. They should, and do, make plant supervisors increase precautions against wrong charging of vessels and incomplete washing out. The official committee procedure, with members looking over their shoulders for popular and press reactions, is to institute specification limits and tests to make sure that the same accident does not happen again. The approach is too stereotyped. While effort is thus concentrated, the next mistake, wholly different, does happen.

There is yet another comment on the possible dangers of dioxin impurities in polychlorophenols. Another product from 2,4,5-trichlorophenol is hexachlorophene, an effective bactericide contained in many proprietary products widely advertised for direct use on the person—particularly the female person. This product has been subject to inquiry as to safety in use for quite different reasons. In none of the many reports concerning the safety of 2,4,5-T has evidence from the pharmaceutical or cosmetic fields been considered, an example of the tendency of committees to be set to work in blinkers and of the prejudice against a pesticide. Hexachlorophene is probably safe, if only because a pharmaceutical product can afford the cost of higher purity standards.

A dilemma

The government committee approach to the safety of pesticides has, as have the pesticides themselves, some unintended side effects. Every new compound believed by a chemical company to be of commercial value must be investigated exhaustively at the company's expense, according to a long list of codified questions. Whenever a new possibility of danger arises, the list is extended. Each new compound has therefore to pass more critical tests than the last. These tests are expensive: the more long-range and subtle, the more expensive. Since simple chemistry has been explored first and most simple compounds are known, each new compound costs more to discover and more to manufacture. Inflation adds to this general escalation. Thus strong pressure is brought to bear on the producing company to develop only those compounds with a wide potential market, at the expense of the speciality product. There is a major effort to maximize sales. Environmental safety is particularly ill served by these tendencies. Specialized products used in careful moderation would be much safer, but no one can afford to prove them so. Some large chemical concerns have already ceased pesticide development.

There appears to be no clear solution to this dilemma. It is hoped, however, that in this chapter some different aspects of the problem have been revealed to readers who have the interest and future responsibilities. The writer can only repeat his plea for a greater understanding by the general public of the nature of the chemical pollution problem. All chemicals must be considered in their continuously changing roles; they must not be condemned by false classification.

FURTHER READING

1. Thompson, T. E. and Bennett, I. (1969), *Science*, **166**, 1532.
2. Binns, W., James, L. F. and Shape, J. L. (1965), in *Symposium on Embryopathic Activity of Drugs*, J. & A. Churchill, London, p. 105.
3. Sarma, P. S. and Padmanaban, G. (1969), in *Toxic Constituents of Plant Foodstuffs*, ed. I. E. Liener, Academic Press, p. 275.
4. Mellanby, K. (1970), *Pesticides and Pollution*, Collins, London, New Naturalist Series.
5. Edwards, C. A. (1970), *Persistent Pesticides in the Environment*, Butterworth, London.
6. Moore, N. W., editor (1966), 'Pesticides in the Environment and their Effect on Wildlife', *J. Appl. Ecol.*, Vol. 3, suppl.

7. 'Residue Reviews'.
8. Kearney, P. C. and Kaufman, D. D., editors (1969), *Degradation of Herbicides*, Dekker, N.Y.
9. De Bach, P., editor (1964), *Biological Control of Insect Pests and Weeds*, Chapman and Hall, London.
10. Hudson, E. J. and Marson, H. W. (1970), 'Role of Phosphates in Eutrophication', *Chemistry and Industry*, 1449.
11. Cadle, R. D. and Allen, E. R. (1970), *Science*, **167**, 243.
12. *Nature*, 1968, **220**, 702: *Chem. Abs.* 1969, **70**, 18509.

CHAPTER FIVE

FOOD, FARMS AND FACTORIES

N. F. Robertson and H. P. Donald

Evolution of food production technology

This chapter describes briefly the epic struggle by man to match his food with his numbers. In this struggle, which he inherited from his most distant ancestors, man has competed with other species, some of whom have lost and died; he has still to settle an internecine war begun long ago. No attempt will be made here to review the voluminous despatches from the many fronts in action today. Readers who can consult the technical journals of the biological and social sciences, as well as the daily and weekly press, will know how great a task that would be. As an alternative, we shall glance at some historical episodes to see how the present state has been reached, and then examine briefly some of the controversial features of the present situation. Our conclusion will be that there are two basic questions facing the human race: (1) the amount of food to be produced; and (2) for how long. Since increasing the first unwisely may decrease the second, man is faced with an agonizing choice. The maintenance of resources and the quality of life command attention in this context; but in the last resort the power to survive will rest with those who have the food.

Domestication

A book could be written about stone-age agricultural technology, for some of it is extant and some has disappeared only in recent times. In New Guinea, Australia, New Zealand and North America, for instance, there are (or were) four different basic types of man, determined by climate and by sources of food. In tropical countries like New Guinea, where the weather is hot and wet, and large animals scarce, forest dwellers fell trees to let in light, plant bananas, keep down weeds for a few years, and then move to another patch; in Australia, aborigines in hot dry areas hunt but do not farm. In temperate New Zealand, in which there were fish but no mammals, Maoris ate birds, fish and cultivated kumara (*Ipomoea batatas*); in North America, Red Indians lived off the natural resources of plants and large animals. In each country, human populations were kept small by disease, periodic starvation, and tribal conflict. In each population, motives and actions in the constant effort to survive were governed by deep convictions about the influence of local spirits and deities.

There is no need to assume either that domestication of animals preceded domestication of plants or that the reverse order was followed. In some areas it might be the first, in some the second, and in some a combination of both. In whatever regions domestication originated, its progress would depend on local conditions. A pre-domestication phase, during which men learnt how to catch, feed, tame and breed wild animals, must have occurred. In some respects, ancient skills would have been inferior by modern standards; in other respects, such as catching and taming, they may at times have been superior. The equivalent phase in plant history might have been the grasping of ideas, such as the sowing of seed and the importance of light, and then applying these ideas to such diverse plants as wheat in the Middle East, maize in North America and quinoa (*Chenopodium quinoa*) in South America. Because they would not have to be sown if imperfectly harvested, plants with tubers may in the first instance have offered a less challenging problem.

In some regions, the evolution of agricultural technology seems to have stopped at this point. The mobility forced on groups of people by exhausted soil or food supplies is represented today by modified nomadism, migration and transhumance. It set strict limits on deviations from traditional routines. Furthermore, such movements had to occur within the territory that the group was able to defend. This principle of territoriality is common to the whole animal kingdom: neither men in general nor farmers in particular fail to obey it.

Where subsequent phases occurred, plants and animals were modified so that they fitted more accurately the purposes for which they were kept. This process of domestication has been continuing for a long time, possibly for ten thousand years in some cases. Figure 5.1 shows how the breeder's work has changed the pig. Its original shape, growth rate, size, fertility and temperament have been replaced by new traits which make it more suitable for bacon and pork. But time is no real measure of the degree of domestication. Elephants, reindeer, camels, mink and some cattle (in Northern India) are not distinguishable from the wild animals which they

Figure 5.1 Fashions in pigs. According to Low's *Domesticated Animals* (1845). 'The first is an outline of the Old English Sow, exhibiting almost all the characters of external form which breeders study to avoid; the second is an outline of a cross between a female of the Siamese race and a native male of a fine breed, shewing the characters which are held to be good, and the consequent tendency to obesity which these characters indicate.' The third is the outline of a Norwegian Landrace boar (bred by the Animal Breeding Research Organization) which showed by far the best result on performance test (1974) of all pigs of all breeds to date.

often rejoin. At the other extreme there are poultry, sheep and dog breeds which, for obvious reasons, would not survive for long in the wild. Through selective breeding, they have lost the ability and the will to exist on their own. Other breeds, and most if not all cattle, pigs and horses will quickly revert or become feral. Horses, cattle, goats and pigs that have become wild are well known.

In dynastic Egypt (2000–3000 B.C.), the five chief uses for animals were for religious ceremonies, pets, food, work and sport. These uses continue but, for completeness, other current uses will be mentioned. The most important of these are for skins, wool and tallow but, in addition, animals sometimes serve as a medium of exchange or barter, and very often as a status symbol. As a rule they fulfil more than one function at a time. Sport and work are at times inseparable for horses and dogs, as are the religious and food uses of cattle and goats. In the circumstances, the attitudes of people to animals are unlikely to be wholly rational. To some extent they will spring from deep-seated human reactions of long standing, and to some extent from conditioning since childhood. Religious discrimination affecting pigs and cattle is well known, as is the popularity of horses, dogs and cats for pets, but not for food. Opinions of agricultural practices with animals vary, depending on whether or not those holding them live in towns, and by which of the five basic interests they are most animated. Although plants also engender emotional responses, they are scarcely of the same order of intensity or variety.

The process of domestication, by which plants have been adapted to human use, has gone further than in animals. The main uses are food, tobacco, rubber, drugs, textiles, timber and amenity. Pride of place in this respect must go to the staple food crops, the cereals: wheat, barley, rice, millet and maize. They have been modified to yield heavily under cultivation and high soil fertility, while at the same time they lend themselves to combine-harvesting by ripening uniformly and resisting wind damage. Another plant of growing popularity in the United States shows how plant breeders can condense the process of domestication from thousands into tens of years. This is buckwheat. Nutritionally superior to most grains, especially in protein and the scarce amino-acid lysine, buckwheat had, unfortunately, the habit of producing new growth and flowers at the same time, thus depressing the yield of grain, It also had weak stems and responded poorly to fertilizer. Good progress is being made towards overcoming these defects. Larger seeds and stronger stems have been achieved, and it is now possible to self-pollinate plants, so that good strains can be preserved and crossed only when planned. Thus will buckwheat be tamed and made amenable to highly capitalized intensive husbandry. Less-visible improvements due to changes in composition and disease resistance can be of equal practical importance. The sugar content of beet and resistance to certain diseases of potatoes are good examples. There is no possibility even of sampling adequately the long list of plant breeding successes, but the horticultural crops and the forest trees must be mentioned. Innumerable new varieties of garden plants and trees have been produced which give pleasure to both town and country people. Although many are abused and mutilated, the mistreatment occasions very little offence. Nor does the mass-production technology of modern horticulture and farming. Both plant and animal domestication aim to produce husbandry systems and the corresponding varieties which will produce food of the quantity and quality demanded. It is inevitable that combinations to which we have grown accustomed will become obsolete, and we shall have to accept new methods, new plants, and new animals still further removed from their original forms.

The agricultural revolution

Down the ages, until the industrial revolution stimulated the growth of cities, the majority of people were concerned with subsistence farming. Progress was slow; so slow that until about 1700 little advance had been made on either the implements or the methods of Greece and Rome (Fussell, 1972). Farmers had become settled on their plots and relied on scratching the soil, applying their hoarded manure and fallowing regularly —a brutish life, always threatened by hunger. As late as 1750, the yield of oats in Scotland was about three times the weight of seed sown: 'ane to gnaw, ane to graw, and ane to pay the laird withal'. Nowadays a farmer would hope for at least a twenty-fold return.

Some surplus above the needs of subsistence has always been available to feed townspeople, most of it no doubt coming from nearby. During the eighteenth century in the United Kingdom, the growth of cities had begun a movement that was to reduce the farming population to a mere 4% or thereabouts of the total and there had to be a corresponding growth of food supplies to the towns. Some supplies were imported, but most were an outcome of the agricultural revolution which began then and has not yet been completed. An agriculture which had met the needs and ambitions of people for two thousand years with little change, was to be intensified almost beyond recognition in the course of a comparatively short 200 years. A readable account of this period in English farming has been given by Ernle (1961). Some recent figures detailing progress are given in Table 5.1 which shows that the United Kingdom output of crop products has risen markedly since the last war. Although all aspects of the industry have been modified, there is much about it still that an ancient Greek would recognize: cultivation, manuring, land tenure and more. We even have the counterpart of those poets who thought it bliss to live in Arcadia—though any farmer in that region who could read would doubtless have wondered why—and they are a species of environmentalist.

Table 5.1 Estimated crop yields in the United Kingdom (Ann. Abstr. Statistics, HMSO)

Crop	*Unit*	*1932*	*1942*	*1952*	*1962*	*1972*
Wheat	cwt/ac	17·4	20·4	22·7	34·7	33·8
Barley	,,	16·2	18·9	2·5	29·0	32·2
Oats	,,	16·3	17·2	19·2	23·1	31·6
Potatoes	tons/ac	6·8	7·2	7·9	9·0	11·0
Sugarbeet	,,	—	9·3	10·4	12·6	13·1

1 cwt/acre = 125·3 kg/ha = 0·125 tonne/ha
1 ton/acre = 2·51 tonnes/ha

This is not the place to set out, however concisely, a chronological sequence of events in British agriculture. For our purposes it is more important to list the chief kinds of development and although referred to separately, these were not and are not independent. In the complex web of circumstance in agricultural production systems, a change in the strength of one strand alters the shape of the whole.

The growth of capitalism and a market economy with their repercussions on road transport, financial transactions, and the status of farm labour was perhaps the most fundamental change. It began with the export of wool, and has reached a point now where prices and quantities of all products are influenced strongly by world conditions of supply and demand. Ultimately these factors affect the adjustments farmers must make to their methods of production if they are to make a living. The same remark applies to another important influence, namely, the nature of land tenure. Down the centuries, several types have been applied in Britain, starting with land holding by strong-arm methods. Enclosures, tenancies, amalgamations and security of tenure were stages by which more and more money was invested in farms with a view to greater outputs from each farm and each worker. Opportunities were created for the use of new designs for machines, and the stage was set for applying the results of the burst of scientific activity. As a result, fertilizers and other advances in chemistry, such as drugs for disease control, weed killers and insecticides, enabled production to rise rapidly. New skills in feeding, better housing and improved varieties of crops and livestock also helped. Where a man sowed, he was almost sure to reap.

Because of its international as well as national significance, one outcome of all this deserves particular notice, namely, the use of concentrates (cereals and high-protein substances, such as dried ground fish meal) for the feeding of livestock.

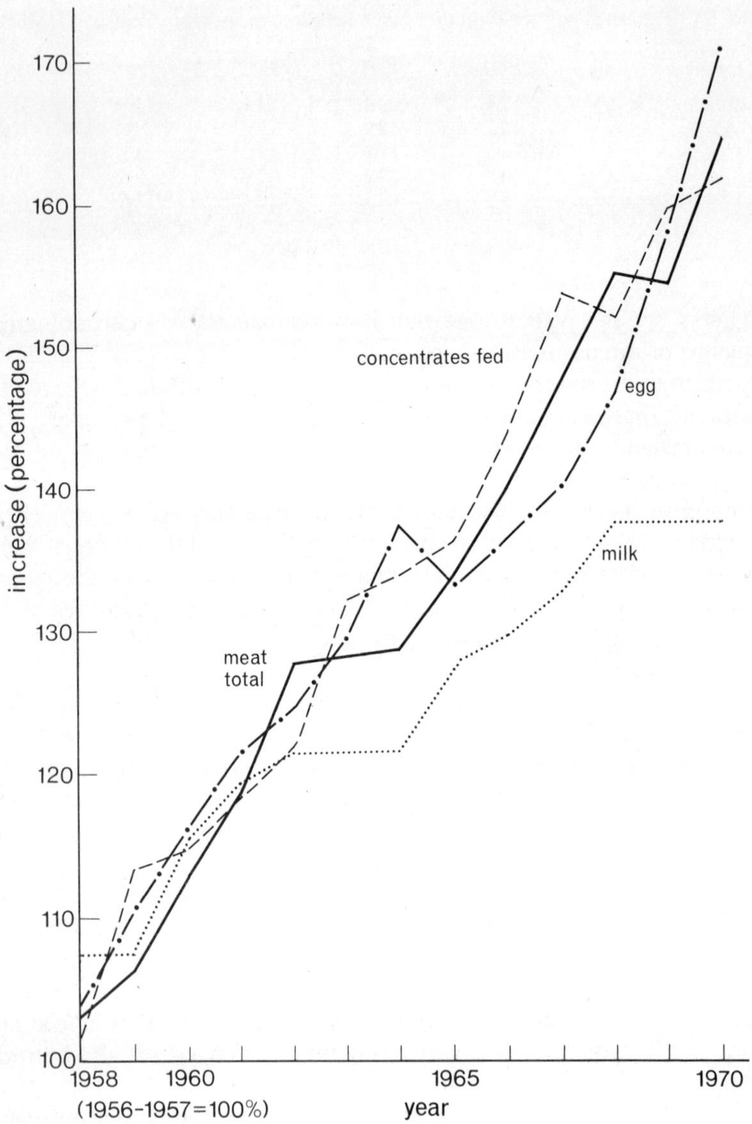

Figure 5.2 Relationship between growth in animal production and amount of concentrates fed in EEC countries (Jasiorowski, 1973).

No doubt recent price increases have altered the trends, but to maintain the levels of home-produced meat, milk and eggs, large quantities of cereals and protein foods for housed animals must be imported by countries that do not grow enough of their own, such as Denmark, the Netherlands and Britain. This is equivalent to leasing arable land. From now on it will be increasingly unlikely that there will be enough grain to feed everyone adequately, as well as to maintain the supply of meat. As figure 5.2 shows, in 1970 farmers in EEC countries were still increasing the amounts of concentrates they fed to livestock, except perhaps dairy cows. From 1956 to 1970, consumption of concentrates rose by 60% and the production of meat by 65%, eggs by 70%, and milk by 40%. The figure also makes plain that animal production in those countries depends so heavily on concentrates that consumers face an awkward and worsening ethical problem.

High grain prices will discourage the use of cereals for livestock, to the extent that consumers reduce the amounts of meat they eat. In wealthy countries accustomed to eating much of it, the reduction may not fall equally on all products, and indeed it may not be very large. In the United Kingdom the degree of self-sufficiency, and hence the need for expensive imports of meat, varies widely with the product. For ten products, the situation as it was before 1939 and as it was estimated to be in 1972/3 is shown in Table 5.2—which provokes the question: What will a column for 1982/3 show?

Table 5.2 Degree of self-sufficiency in agricultural products in the United Kingdom. Total production shown as percentage of total supply on the home market (OECD, 1974)

	Before 1939	*1972/73 (estimated)*
Wheat	23	52
Barley	46	95
Butter	9	22
Cheese	24	54
Eggs	61	97
Beef and veal	49	85
Mutton and lamb	36	43
Bacon and ham	29	44
Pork	28	93
Poultry meat	80	99

Restructuring agriculture

Since about 1950 three other developments have occurred to emphasize the breakaway from the traditional rotation of crops, uncertain prices and small unorganized farms: (1) action by government to influence prices, activate research, and control disease; (2) increasing attention to quality and marketing; (3) organizations of various kinds to carry out large-scale operations. Since there is a great diversity of structure in all three categories, it may be reasonable to infer that it is not yet clear which are the most suitable devices for the several purposes. The production of broilers and eggs is increasingly dominated by a few very large producers; the production of milk and lamb and wool by many small ones. The marketing of milk is done by a few quasi-public boards of producers who also take care of quality control and provide artificial-insemination services, but much of the marketing of meat is in the hands of limited liability companies.

On the assumption that the demand for food as cheap as is possible will continue and grow, it will be difficult to resist the adoption of large-scale industrial methods if they prove better able to meet the demand than small-scale methods—but they may not be the same as those found by trial and error in poultry production. Dry-lot feeding of purchased beef cattle in the United States, for instance, developed scale, advanced technology in feeding, buying and marketing, but no buildings to house the livestock. In fact in this and in other examples, such as bacon pig production, there may be little to distinguish the large organization from a farm, except its size, and the specialization with centralized financial control and decision-making. Very large pig and dairy farms are now scarcely newsworthy, but so far only the pioneering stage in large-scale sheep, rabbit and fish farming has been reached.

There is of course nothing peculiar to agriculture in all this. Specialization, scale, advanced technology, and the concentrating of capital and labour into fewer units, has recently (1974) been urged by the National Economic Development Council for British industry generally. Changes in agriculture, however, generate their own side-effects on the industry. On this complex subject all that can be said here is that, as well as structural changes, the trend to intensification involves departures from traditional attitudes to farming. Farming is characteristically small-scale, individualistic, variable, adaptable, vocational. Industrial food production has none of these attributes. To approach the problems of agriculture, whether they be environmental, social, scientific, economic or political, as if farming were still in its pre-war state is likely to be unprofitable.

Technological imperatives

Those who have studied the development of modern industrial technology (Korach, 1964) have come to recognize certain laws which govern the relationships of the factors combining to produce the technology. It is interesting to examine these laws to discover whether they are applicable to the agricultural technology which concerns us here.

The *law of the cost-variable* states that each process has a maximum allowable cost determined by the market price of the product, and that no technology can be applied if its product cost is higher than this. This law is basically applicable to agricultural technology, but the time scale which obtains, for example, in the production of meat means that this law is often broken by the market swings between the time when the product is planned and the time when it is sold. It is the aim of all agricultural policy and planning to ensure that the law is not broken. When it is, there is hardship in the agricultural sector.

The *law of the great number of variables* states that in technology only a limited number of variables, the dominating ones, can be considered; therefore technology will constantly seek to simplify and reduce variables. In agricultural technology the effort required to simplify, in the face of biological variability and the variability imposed by climate and weather, must be even greater. In this law can be seen the reason for the move towards environmental control and housing of animals, because of the simplification which gives an immediate boost to technological efficiency.

The *law of the scale effect* states that at a certain degree of quantitative change in production a qualitative change becomes necessary. This is well seen in the increasing use of machinery to handle rising output. Combine-harvesting and milking parlours are good examples. Another is the huge incubator which makes the production of millions of chicks possible in a short time, and is essential to the organization of the modern broiler house. Still more mechanization to replace human muscles, and new ways of saving time are in train for animal husbandry. As sheep fertility rises, smaller lambs born as twins and triplets need more care. As concentrations of cattle grow, mechanical food distribution becomes essential.

The *law of automatization*, stating that the range of variability can be held between determined limits only in automatic processes, is clearly broken by the products of much agricultural technology; but such automatization as is possible in controlling environment and food supply leads to a reduction in variability.

We can see in these four laws a pressure from the market for agricultural products for (*a*) the cheapening of inputs, (*b*) the simplification of production processes, (*c*) a rising scale of production, and (*d*) the automatization of processes. All these are pressures for movement away from farming as we have come to accept it, i.e. mixed stock and arable farming on a relatively limited scale. But in a world where food and cheap food will be even more in demand, it seems inevitable that farming in the future will have to learn to obey the laws of general technology.

It is not appropriate here to discuss Korach's other laws of development, interesting as they are, for an understanding of agricultural technology.

But it might be useful to point to the importance of *model building* in the development of industrial technologies. The development of a process on a small laboratory scale is worked up through a series of pilot plants or models until full-scale production is achieved. Such a logical progression is not so applicable in agricultural technology where the units of production are to some degree already determined as individual animals or fields. But an important facet of modern agricultural research is the development of systems by the breaking down of production patterns into a series of small-scale enterprises which can then be resynthesized into a new system. Since such model analyses with live animals and crops are very expensive, model building using computer techniques takes its place. But all this work is designed to simplify, to cheapen, and to encourage enterprises to grow in efficiency.

Industrialization

Forms of production

It is easy to see what is meant by industrial forms of production by comparing them with a handicraft. We can think of woollen manufacture evolving through a simple hand-spinning process to the spinning wheel, and then by multiplication and mechanical ingenuity becoming a large-scale production, and at that point becoming industrialized. Industrial production, however, does not emerge until there is a relatively large market for the product. Its prime characteristics are the large scale of the manufacture, the relatively low labour input, and the high capital cost of the machinery. One common consequence is that the number of products is reduced, so that most advantage can be taken of a valuable machine; but in agriculture there is greater difficulty in identifying the transition point between the hand or craft industry and the industrial process. As soon as man began to sow seed, he had begun to increase the scale of his operations and to decrease the labour input for a given level of production. The associated increased capital input might be said in this instance to be represented by the seed held over from year to year. Indeed, we might go further and identify a form of organizational industrialization in Roman Britain, where the success of the Roman Villa settlements is said to have rested on a skilled organization of a number of small units of production rather than on the scale of production of any one unit. The same difficulties obtain for the identification of industrialized animal agriculture. The non-industrial situation would rest with a single cow and a few sheep, but the very nature of cattle which naturally form herds and the characteristic flocking of sheep, together with the extensive system of keeping flocks and herds in a nomadic situation, means that the cow and the sheep would naturally lend themselves to large-scale management.

Is the nomadic shepherd of the Middle East a fore-runner of the industrialist of to-day?

These relatively simple and unsophisticated forms of industrialization, however, are not the ones usually in mind. Industrialization has a connotation of unconformity with the environment, and seems to imply a situation where the enterprise is larger than the environment can comfortably contain, so that it noticeably disrupts it. It can do this in various ways—socially, economically, or environmentally, and possibly in all three.

Industrialized agriculture undoubtedly exists when measured in terms of market size, scale or production, limitation of product variety, and relatively low labour input. Such industrial developments of agriculture occur in a number of ecological situations and under different political regimes.

The simplest form of animal agriculture, the extensive ranching of cattle or sheep over an area with natural vegetation, shows unconformity when the grazing or browsing pressure is such that rapid change takes place in the natural vegetation; it is then degraded in terms of the plant stature (i.e. trees and shrubs are killed out) or in terms of the numbers of plant species. The extreme situation is reached in areas where over-grazing, together with excessive trampling, leads to soil erosion and the loss of a natural vegetational cover. Sometimes, however, an attractive landscape is also the result of industrialized ranching. The appearance of some Scottish hills is due to the grass cover (mainly *Nardus stricta* and *Festuca ovina*) on the less depleted soils; the appearance of other Scottish hills is due to the heather cover (*Calluna vulgaris*) which follows burning and grazing on the more depleted soils. They are in fact products of industrial unconformity. And some of the rare plants of open habitats, e.g. *Primula farinosa*, are found where the pasture is kept open by pressure of cattle grazing.

In dry climates such as that of California, and by adaptations in the less climatically favoured states of the United States, another comparatively simple system of animal production is developed. This is the *dry lot* or *feed lot* in which young cattle are penned and fed so that they grow and fatten. It is the scale rather than the nature of this operation that leads to its unconformity with the surroundings. But it also produces environmental problems. These are due to air and water pollution, and to the crowding of the cattle. However, what would be totally unacceptable in the wetter maritime climate of the United Kingdom becomes bearable under dry-climate conditions. There is no doubt, however, that the sheer scale of a dry lot with as many as 20,000 cattle at a single location gives the enterprise a factory-like quality to human eyes; this is intensified when, as is usual, the feed for the animals is milled and compounded at the same location. If abattoir facilities are also adjacent, the resemblance to a factory is very close, since an essential feature of a factory is a building.

Industrialized animal production is seen at its fullest development in intensive poultry production, where large-scale developments either for egg production, or for broiler or turkey production, become largely self-contained, the eggs for hatching being as a rule brought in from outside, together with the raw materials for feed production. Animals move through the factory in batches and are kept in a strictly controlled atmosphere and crowded conditions.

The various industrialized farming methods which have been mentioned are spread widely throughout the world. Ranching is found in large areas of North America, and on the grasslands of South America where cattle production is highly developed. In both situations, production ends in the feed lot, the cows being ranched and the half-grown progeny being confined for fattening. Similar cattle ranching takes place in parts of Australia but, there, sheep husbandry has been developed on an extensive ranching system as well. The wide variety of climates in the USSR and in the adjacent countries of the Communist Bloc show a great variety of animal agriculture, but the intensive housing of cattle, sheep and pigs in association with grain feeding, particularly of maize, is a development which, at least until recently, was increasing.

No aspect of livestock production is incapable of industrialization. Pigs, because of their largely cereal diet and their favourable response to controlled environment, lend themselves to it. Cattle and sheep are also more and more being confined (at least at the fattening stage) in feed lots or buildings where a measure of environmental control and management control can be exercised.

In plant agriculture, cereal crops are thoroughly industrialized. As a result, the growing of a single type of crop over many square miles brings an unconformity with the environment. This is particularly noticeable with the monoculture of trees (whether fruit or timber) and cereals. In the home of the potato in South America, the beginnings of tuber agriculture can be seen in scattered patches of potatoes, often of mixed varieties and even with more than one species. Tubers which are left after harvest form the seed of the following year. Compare this with modern potato production, where specialized crops are set aside for seed production, where health regulations for seed production are stringent, where cultivation forms a very large part of the cost of crop production, and where spraying is routine and essential for a healthy crop. Following harvest by complex machines, now even with electronic sorting devices for sorting potatoes from stones, there is careful storage in special buildings. The same picture can be found in fruit production. Here the single tree or group of bushes of earlier times has been replaced by orchards, which in some countries cover many square miles and are subjected to complex pruning operations and stringent spraying regimes, often repeated up to ten or a dozen times in a growing season, in order to produce unblemished fruit for the market. Is this to be considered industralized production following the criteria adopted at the beginning of this section? Modern crop production fits all the criteria. It has arisen in response to large market needs and it has developed size with centralized financial control. The number of products (in this case varieties) has been reduced, and the management is directed towards a relatively low labour input and high capital input. Moreover, it shows an unconformity with its environment. The natural vegetation is excluded, and a single crop takes over large areas of countryside; yet it could be hard to convince people unconnected with agriculture that such crop production techniques represent an industrialized process. Indeed some people actually derive aesthetic pleasure from contemplating crop production on this scale. The regular rows, the changing facies throughout the year, as young crop gives rise to mature crop, as seed bed gives rise to harvest, are accepted as part of the landscape and please a great many people. In these circumstances it may be questioned whether the distaste which many people feel for modern industrialized animal production is not a temporary distaste which, as a new generation becomes accustomed to large-scale production, will become converted to a form of aesthetic satisfaction.

Adaptation of plants and animals

When industrialization takes place in agriculture, the species and varieties which have arisen under domestication have to be further adapted to fit in with the industrialized process. This is most readily seen in crop production where, for example, apple varieties which are highly valued for their dessert quality, for their aroma, or for their cooking quality, are dismissed from the industrial scene because of characteristics which interfere with production processes. For example, the early eating apple *James Grieve* has no industrial value because of its extreme susceptibility to bruising; and the old culinary variety *Stirling Castle*, which forms a veritable 'apple snow' on simple boiling, was immediately excluded from commercial development because of its susceptibility to sulphur-containing fungicides. Where breeding is sufficiently rapid to allow of adaptation to industrial needs, as with the cereal crop, there is a concomitant development of yield, of capacity to respond to high fertilizer inputs and the development of shorter and stiffer straws. The potato is a splendid example of the breeder's capacity to develop varieties responsive to intensive cropping.

The large-scale growing of potatoes, which is perhaps only 200 years old in Europe, brought with it the dangers of epidemic plant disease, so well illustrated by the potato-blight-induced famines of the 1840s. The effort which was devoted to the breeding of varieties of potato resistant to blight, while unsuccessful in itself, increased the scale of operations and caused breeders to consider a number of other characteristics. Tubers free from deep eyes, tubers which resisted bruising in the harvesting operation, and plants which held their tubers close to the stem base to facilitate harvesting, were sought in new varieties. Other impressive examples of this adaptive breeding are to be found in raspberries and tomatoes.

The process of domestication clearly requires the selection of animals which show a high degree of tractability, otherwise they could not be handled expeditiously by herdsmen. Modern strains of poultry are the result of selecting animals which are capable of remaining productive under the conditions of the intensive laying house. Likewise the shape and composition of the pig has been adapted to the requirements of the bacon industry.

It has been suggested that people often identify themselves with animals, but not with plants, and it may be for this reason that no-one objects to 'improvements' on Nature's handiwork, such as the cauliflower (which is a condensed inflorescence carried to monstrous extremes) or a prize pumpkin. Provided there has been time to get used to the new form, however, the same applies to equally extravagant modifications of animals. Bull-dogs, for instance, that need a midwife when born, or Pietrain pigs, that are prone to drop dead with fright, or double-breasted turkeys that have difficulty in mating, are acceptable, whereas a wingless featherless broiler suddenly introduced to commerce would attract comment.

Mechanization

Implicit in the industrialization of agriculture is the mechanizing of production. This leads to a change in the agricultural scenery, to a change in the conditions under which crops and animals are produced, and to developments in scale which tend to remove the older style of farming from the view of ordinary people. In days gone by anyone driving through the countryside could observe the agricultural scene, and even today there is much of interest to be seen. But as crop acreages become greater, and as animals become more confined, there will be less variety to be observed. In another part of this chapter the question will be implied: Do we require our agriculture to contribute to our environment? Where mechanization is applicable to crops, large-scale cultivation obviously simplifies and speeds up the work of the farmer. The development of large powered tools enables work to be done quickly and to best effect over large acreages. The management of crop production therefore becomes much easier. In animal husbandry the mechanization of feeding removes a very laborious chore. It is already advanced in the poultry industry, widespread in the pig industry, and developing in relation to housed fattening cattle.

The removal of dung is a major problem in modern farming which is met in various ways. In modern poultry production this is often highly mechanized, with continuous belts removing the faeces to a dropping pit, where they can be removed by tractor shovel. In pig and cattle production the excreta are often disposed of by way of slatted floors, which allow the material to accumulate. Later it can be disposed of by pumping it into a slurry tank and dispersing it over the land. In other cattle systems the animals may be bedded on straw, which is used to contain the excreta and is afterwards disposed of as dung by rather laborious tractor shovelling; or yet again the passageway may be continuously scraped by a mechanical scraper blade which directs the slurry into a pit, whence it is removed either by tanker or by a system of pipes to the pastures. As enterprises grow, it is likely that they will require to develop sewage treatment works associated with their slurry disposal. In future factory production there will be large feed stores, and feed-mill-and-mix units handling concentrated rations, sometimes with associated silage or other forage. The egested remnants will be processed at the other end of the building by a sewage-disposal unit. Stockmen in this situation will act the role of the factory supervisor with a very clear view of the standards of health required, but with little close care of individual animals. This means a change in the traditional relationship between man and his livestock.

The capital involved in the development of large-scale crop production and particularly large-scale animal production units will clearly be considerable. At the moment the cost of housing suckler cows in relatively simple buildings is calculated at £60–£100 per cow on 1974 figures. This sum might be increased fivefold by the provision of sophisticated feeding and slurry disposal equipment. Production units will have to grow in size to develop the maximum economy of scale. This means that the largest number of animals must be fed by the equipment available and their residues disposed of to make maximum use of the money invested in sewage treatment works.

Organization

Expenditure of this order moves farming out of its present financial structure. In Britain, about half the farms in the country are owner-occupied and extend to 200–600 acres, which provide a reasonable living for one man and his family. In terms of capital return, the owning of land has to date provided only a small return on total capital invested—

something of the order of 2–4% of the total investment. In tenanted farms, the land is rented from a landlord who is content to receive a low return on his invested capital, partly because of estate-duty relief and of the capital appreciation he has come to expect. The tenant farmer contributes tenant's capital, which yields a return similar to the return on capital in any other business, that is to say anything between 10% and 25% gross yield on invested capital before allowing for management costs. Developments in mechanization and scale of enterprise in the future will demand capital investments quite beyond the capacity of the present landlord or tenant to provide. Farmers will either combine in co-operative ventures to develop high-cost enterprises, or money from general investment sources will be channelled into agriculture. For example, feed companies may take part in a form of vertical integration with farmers who in return for an investment of food for their livestock will commit themselves to supply at contract prices the products of their enterprises.

Industrialization of agriculture is closely tied to technological progress in other fields; and almost inevitably changes in agriculture will result from advances in adjacent industries. For example, the development of chemical technology for the production of amino-acids and feed additives makes feeding regimes possible which had previously been inhibited. Similarly, the construction of buildings for livestock was a major impediment when houses were built of stone or of brick; but modern methods, which allow the development of wide-span buildings built in light durable materials at comparatively low cost, immediately open up prospects for the expansion of intensive animal husbandry which did not previously exist. Such possibilities raise aesthetic problems in the countryside, because large blocks of buildings begin to dominate the landscape.

The key to a number of past and present developments in animal production lies in disease control. When there was no understanding of cattle scours and no treatment for calf pneumonia, rearing calves in large groups was very risky. Modern veterinary medicine, however, has allowed (a sometimes uneasy) development of intensive calf production.

Again, a growing business in earlier days was often hampered by its inability to keep its accounts, stock records and so on, but modern computer technology makes such a barrier no longer important. Computerization has moved into other fields too. The breeding policy for laying hens, for example, is now determined by the study of many matings; and the most economical mixtures of feedstuffs worked out. The poultry companies have gone much further. Some are now multi-national; and extension of this idea to the point where the natural advantages of particular parts of the world can be harnessed together by international organizations handling other animals cannot be far away. In a country like Britain, which has large areas of mountain and hill land, sheep and cattle will be required to exploit herbage which grows in those areas and which cannot be harvested otherwise. Animals from them form a breeding base for the production of fattening animals. In the sheep industry, there is a geographical stratification such that, on the higher hills, hardy breeds like the Blackface or Swaledale are bred. Crosses with other breeds, such as the Border Leicester, produce Greyface or Masham ewes which are then crossed with a Down ram to give rise to lambs for fat-lamb production. At the moment, the final lamb products of this system are fattened off their mothers, or after a short spell of supplementary feeding in the early winter. But animal technology may well advance to the point where the ewes can be encouraged to breed up to twice in any one year, and a corresponding technology may develop which will allow lambs to be weaned at an early stage and intensively fattened. The market price of the end product will, of course, have to cover the costs. The situation in cattle also lends itself to further intensification. Here there is the added complication (and opportunity) that dairy herds may, and often do, produce progeny which are used for beef. When it becomes possible to arrange the sex of a calf at the time of mating, the whole structure of the industry will be further rationalized.

Although the collecting of large numbers of young animals under one roof for fattening is merely an extension of an old principle, the use of the term *factory farming* brings with it emotional overtones, and shifts thoughts from the physical to the metaphysical. Other modifications to agricultural technology also bring emotional reactions, and consequently the rate of change becomes very important. Earlier it was mentioned that society accepts large acreages of cereals or intensively grown crops as part of the scenery, but as yet is undecided whether to take offence at large aggregations of animals, particularly if they are housed. If such changes can be absorbed at a rate which allows the human mind to adjust, then it may be that they will be accepted without too much difficulty.

What is important in this situation is for the community at large to identify its objectives. The agriculturalist assumes at present that he should try to produce food as economically as possible and, provided no overt cruelty is involved, he must take advantage of every technological development. If modern poultry production methods were set aside, egg supplies would become seasonally scarce, quality uneven, and price very much higher. Since consumers are sensitive to price and reduce their purchases when prices rise, it follows that most of them are not prepared at present to do without the benefits of intensification. These benefits, however, cannot be expected to cancel all the effects of rising costs of production and so the UK consumer must get used to higher food costs.

Social aspects

The problems

Table 5.3 has been compiled from an OECD publication (1974) to give some basic facts about agriculture in the United Kingdom. Obviously agriculture cannot be thought of by urban people as a source of food and nothing more. One quarter of the population lives in or close to the country and 0·7 million of them earn most or all of their living there. Only 2·7% of the gross national product comes from agriculture, but that includes about 75% by value of the food supply. The remainder is imported. As transport has become more readily available to everyone, the notion that the countryside is part of the environment and one determinant of the quality of life has sunk in deeply. It is not to be ravished or squandered.

Table 5.3 Aspects of UK agricultural industry 1972 (OECD, 1974)

Land	Agricultural area	19·1 million ha
People	Total	55·8 million
	rural population	12·9 million
Work force	Total	25·4 million
	agricultural: full-time	0·5 million
	part-time	0·2 million
Production	Gross domestic product	£53,848 million
	Gross agricultural product	£1,461 million
	as %	2·7
Imports	Total value	£11,155 million
	agricultural products	£2,767 million
	as %	24·8

To a remarkable degree, the interest of the public in agriculture seems to arise less from fear of energy costs or food shortage than from concern about amenity, pollution, and animal welfare. Farming is liable to conjure up thoughts of silent springs, dead fish, and cruelty. However unjust this may be, it is for the farming industry and the research and development organization which serves it, to devote some of their resources to learning how to preserve national pride in the countryside, and how to cope with weeds and pests and livestock acceptably. Another obvious subject is provided by aspects of factory production, such as animal welfare and air pollution. It is not merely a question of being sensitive to public criticisms of agricultural practices where they are deserved. It is equally a question of knowing how to counter ill-founded complaints, and how to anticipate them by reference to a solid body of information. Not so very far back in history, agriculture supported only a primitive national life and resisted change. Now that it is no longer frozen by a rigid set of defence mechanisms, such as fixed rotations and easily terminated tenancies, it can be made a living creating culture, capable of keeping pace with society—

not peripheral to it, but as central as eating and loving. The arguments about the EEC, the CAP and about the direction of R and D, are just parts of the incessant dialectic between creative man and those representing the power of current social and economic forces.

An ethical dilemma

Scientists trying to raise food production or reduce population growth rates face a dilemma. Concerning medical success in reducing death rates, Professor A. V. Hill said:

> Suppose it were certain now that the pressure of increasing population, uncontrolled by disease, would lead to not only widespread exhaustion of the soil and of other capital resources, but also to continuing and increasing international tension and disorder, making it hard for civilization itself to survive, would the majority of humans and reasonable people then change their minds about controlling disease? If ethical principles deny our right to do evil in order that good may come, are we justified in doing good when the forseeable consequence is evil?

This is no imaginary dilemma for agriculturalists. A corollary of Malthus' theory called the 'utterly dismal theorem' has been proposed which fits their case. Any technological improvement, when the only check on the growth of population is starvation and misery, will have the ultimate effect of increasing the sum of human misery because a larger population will experience it. The world can avoid this result by falsifying the premise, i.e. by finding ways of raising food production and stopping waste, while devising some less disagreeable ways of limiting the numbers of mouths. Many writers regard the task as hopeless within the time available. William and Paul Paddock (1968) for instance in a book called *Famine—1975!* estimated that world food supplies would need to increase by 26% within ten years, a highly unlikely rate. Research and development however will not be stopped while this controversy goes on.

The dynamic of chance

In order to create a basis on which to found a point of view about the effects of technological development, some assumptions need to be made. For present purposes these will be:

(*a*) Demand for food will rise with world population. There is an alternative which is that the world will experience substantial periodical reductions in numbers like the St Kilda sheep.

(*b*) There will be a slow trend to more technology. A revolutionary change in land tenure, for example to collectives, would not affect this much. Adaptation would be slow then, too, on a national scale because of the complexity of any system of food production. The ultimate objective of agricultural research and development is to release man from some of the uncertainties associated with outdoor farming.

(*c*) There will be a slow substitution of animal products by both natural and artificial plant products. This will not happen evenly to all countries.

(*d*) The mix or proportions of land, labour, and capital, in the organization of food production will vary with national resources of these components, and therefore the most effective technology will likewise tend to vary.

(*e*) The political urge to be self-sufficient will grow, especially in the poorer nations or in those becoming relatively poorer. Apart from better food supplies, there will be advantages in providing work and in avoiding outlays of foreign exchange.

(*f*) Innovations will as heretofore either die young or follow an S-shaped curve of growth in the extent to which they are adopted.

(*g*) There will tend to be a rising average size of holding in order to promote financing, marketing, and use of technical advances and skilled labour.

These assumptions can be summed up as recognition of a dynamic which produces intensified plant and animal husbandry at all levels. Incapable of it unless subsidized, the lowest level, nomadic pastoral systems, may disappear, but in spite of that, the difference between the highest and lowest will grow. Since there are many production systems and many interrelated aspects of each that are susceptible to improvement, there is a vast number of ways in which the trend to intensify can show itself. Nevertheless, because the interrelationships between plants, animals and people are so strong, all three have to be adapted together, except on specialist enterprises where this does not apply. Naturally enough, technological progress will be faster on them, and technologists will be encouraged to put their efforts in that direction. But that is not all. Society must also decide whether it wants the food and what restrictions, if any, it wishes to place on the methods by which it is won.

Animal welfare

The five chief uses made of animals by man have already been mentioned. All arise from quite basic human needs—food, pets, religion, work and sport—and so are not going to disappear suddenly. Their importance varies from one country to another, and it is relevant to the purpose of this chapter to add that as science-based technology increasingly influences animal and plant husbandry, the significance of religious and work aspects declines. Also, where tractors are common, work horses become uncommon. Judging from the past, it seems likely that the public interest in farm animals will express itself as a series of unpredictable restrictions of minor economic importance. These will be acceptable to farmers. There is perhaps some danger of the public delaying the adoption of new ideas which may come into being only after pressure from the farmers, but there is little danger of adversely affecting the economic health of the livestock industry. Inaccurate though it must be, the balance which weighs the public interest for or against a particular course of action is held by parliamentarians and government officials, and they must reach decisions whether or not there is enough evidence on which to found them. How important it becomes, therefore, that the public should not allow anything in farming that is merely aberrant to become symbolic of farming.

It is against this background that some uneasiness arose about the welfare of farm animals, especially after the publication of the book by Rachel Carson called *Silent Spring*, and another by Ruth Harrison called *Animal Machines*. At various times, protests have led to the banning of badger-baiting and cockfighting; also to the control of export of live animals, and of slaughterhouse methods. A committee was set up, with Professor F. W. Brambell as Chairman, which reported in 1965 on the general topic of animal welfare under intensive systems of production.

The report illustrates very well the semantic, logical and ethical difficulties inherent in the subject of animal welfare. Practices such as meat eating are so well established that the rearing of animals in order to kill and eat them has to be accepted, no matter what the unfortunate consequences may be to the animals. Since the Committee could not very well recommend the abolition of meat eating, they had to accept the need for slaughter houses. Obliged to swallow camels they strained at gnats, as indeed do all carnivorous people who object to both real and fancied cruelties. Cruelty to animals does occur, but finding suitable ways to prevent it proves difficult when it takes a mental rather than a physical form, the degree of which is usually impossible to assess. Since the breeding of strains of animals adapted to close confinement is a feature of modern technology, it is important (without good cause) not to attribute to animals those feelings about confinement which would be felt by human beings in that condition. The evidence from the animals is that they do well what is expected of them, from the point of view of food production, but in respect of their feelings the evidence is equivocal. Very often animals appear to prefer their quarters to their liberty. Whether or not it is ethical to breed animals that do not resent being caged is a question that underlies much discussion of intensive animal husbandry.

The Brambell Committee had difficulty in defining *intensive* and *factory farming*. For its purposes the Committee decided that the term *intensive* would be used for production methods which involved keeping livestock under cover. Obviously this will not do for the broader questions being discussed in this chapter. Here it will mean that a certain production system uses a comparatively large amount of capital or labour on each hectare of land. Rice growing in Japan is labour-intensive, but capital-intensive in the United States. Poultry production in Scotland is capital-intensive while hill sheep farming is not.

The term *factory farming* also caused the Brambell Committee some trouble. It was taken to mean production methods characterized by a large scale of operations and a high degree of automation. This is perhaps as close as we are likely to get to the meaning of a pejorative which is conveniently vague. Since neither farms and farming nor automation are necessarily involved in the methods, however, the term will be replaced here by *factory production*.

After much careful study the members were faced with an old and intractable problem. It is one thing for an individual, possibly an animal lover or a vegetarian, to urge that livestock should be treated with the utmost kindness, comfort and liberty, and it is quite another for a responsible governmental committee to compare and weigh up economic, scientific and emotional factors operating in an important industry. Although they felt unhappy about several practices, the members were handicapped by a lack of objective evidence about the welfare of the animals concerned. Some farm practices are unpleasant, e.g. castration, tail docking, early weaning, but they have weighty practical justification. Others present a mixture of good and bad elements. Close confinement in summer may be a hardship, but in winter a piece of good fortune. Predators and certain diseases are avoided, but other diseases incurred. Some forms of pig housing attract criticism, but they are nothing like as obscene as the apparatus of the abattoir.

The term *factory farming* seems generally intended to convey the idea that large numbers of animals are closely confined and very well fed, so as to shorten their lives or raise their output. Such determined exploitation, notwithstanding the benefits of disease control, and temperature control, is felt by many to be offensive. Since exactly the same objectives have been sought on small farming enterprises for a long time, although often less successfully, it looks as if it is the size of an enterprise which draws attention to it. Those who feel strongly about this are not mollified by economic success. By no means every aspect of the technology is admirable, or is known with certainty to be necessary or unavoidable. Further research may reveal ways of avoiding mutilation and discouraging vices in pigs and poultry, just as it has removed some of the major nutritional problems of housed sheep and cattle.

From its very beginnings domestication interfered with the struggle for existence of plants and animals. Competition was eliminated or changed in character. Animals were prevented from winning and keeping their hunting or grazing territory, and plants were no longer restricted to special ecological niches. Gradually their natural habits have been eroded, and their social relationships disrupted. This has gone farthest where genetic selection has produced types adapted to specific purposes. The remarkable specializations of dogs, such as greyhounds, sheep dogs, miniature poodles and many more, offer perhaps the best examples. From the point of view of food production under intensive conditions, the most important technical achievements have no doubt been the strains of poultry and associated feeding and health practices as used in modern establishments. Biologically efficient, egg layers produce in a year about 240 eggs of the size and colour wanted by the market, and do not go broody and stop laying. Broiler strains of chickens and turkeys have also been bred to very demanding standards of reliability and uniformity. They are not intended for farmyard use, but for factory-style production of large quantities of a standard article of food.

One characteristic of wild animals which has not yet been eliminated from domestic livestock is the urge to establish a place in the peck order of a group. All types of livestock show it still. When animals are loose in a field, they soon discover their place and keep out of the way of superiors; but when closely confined they cannot do so. The result is often painful for the inferior which may have to be rescued.

Pollution

Concentrations of animals inevitably create a risk of air and water pollution. This was true long before the present levels of intensification were reached. Farm smells were once regarded as a normal feature of livestock farming, but there is less willingness now on the part of urban neighbours to put up with them. Regulations governing the disposal of dung and foul drainage-water from cowsheds and silage pits, to prevent it reaching underground water supplies or rivers, have placed a very heavy financial burden on many farmers. They have, however, been necessary to prevent a degree of contamination of air and water which no one would want.

Pollution is an evocative word and is often used to create impressions rather than to sum up facts. Since it is a law of physics that energy is conserved and not destroyed in a transformation, it follows that all effort and all life results in by-products, some of which can be regarded as pollutants, even if apparently harmless, such as heat or CO_2; and the greater the activity the greater the pollution. Although it would be convenient to have a definition of the degree of pollution which by common consent amounted to a nuisance or worse, it is impossible to devise one. Arbitrary standards can be set, but there is no way of measuring the net effects of factory chimneys on the yields of crops, especially glasshouse crops; or the air and water pollution on farms due to fertilizers, silage or animals. It has been said that environment has no price tag, hence the problem of extracting the cost of putting right or preventing the damage from those who cause it. In some cases, environment has in fact a price tag. The locality in which a house stands makes a difference to its value. Next to a piggery its value would be reduced. Other farm enterprises of national importance, such as a large poultry business, may create local 'disamenities'. To what extent such an enterprise should be obliged to pay for these disamenities when it is desirable to have the work and the product, is a problem to which as yet there is no generally accepted solution. If the world food deficit grows, however, it is conceivable that some attitudes may change. There might be freer use of chemicals that are now discouraged for various reasons, e.g. DDT and aldrin for pest control, and penicillin in pig food; and there might be less concern to preserve a familiar landscape. Food production will be encouraged, at a cost to the environment if necessary, unless some effective research achieves the same ends without the cost to the environment.

One more aspect of the pollution problem on farms must be mentioned. It is much quicker to ban some practice from fear of possible consequences, than it is to find out whether or not the fear is well founded—and, if so, to develop an alternative. At present objective data not only about pollution but also about animal welfare and land use are frequently unavailable when questions arise. Agricultural research and development might in the circumstances usefully be extended so that such data are accumulated and made public.

Amenity

Food production and an opportunity for people to earn a living are by no means the only functions of agricultural land. In some areas they may not even be the main ones. The alternative uses in fact cause much dissension, so that many and varied pressure groups emerge. Good arguments can be advanced for mining, forestry, tourism, water catchment, roads and airports, wild-life and recreation, as well as the preservation of well-loved landscapes. All are served by eloquent protagonists. It can hardly be a matter for surprise that there is rarely if ever a firm national policy to guide administrators in preferring some uses to others in particular areas. What is certain is that in developed and undeveloped countries those who live in towns have a very close interest in both food and amenity, and they have political power to look after it. But they can neither preserve the countryside unchanged, nor free themselves of the danger of permanently damaging its ecological stability through over-exploitation. Because the effects of erosion, floods, fires and poisoning can be felt far beyond the country where they occur, concern is also felt for the ecological errors of all nations. Scientific research has done much to point out dangers and devise remedies, and will do more; but the fondest guardians of the countryside are those who live in it. Their prosperity is the best assurance that this national resource will be well cared for and handed on to their successors in good condition.

Much that is treasured in the countryside and landscapes of Britain today was made possible by a social structure that no longer exists. Each generation is at once the creator and the result of the environment in which it lives, and no amount of nostalgia will prevent it leaving its stamp there. What may reasonably be asked by posterity of the present generation is that such changes as are made should add if possible to the quality of life, and certainly not diminish it by encouraging rural slums. If there must be factories, let them be well designed and well sited, wherever they are. If there must be odours, let the offence be minimized. If there must be chemical warfare against weeds and pests, let the risks of breeding resistant types and of unintended damage to other plants and animals be not rashly incurred. While most people would no doubt subscribe to these proposals, it is no easy matter to establish rights to the enjoyment of the environment, and to move thence to the duties relating to them. Owners and occupiers of land are not free to pollute streams, shoot protected birds, or ill-treat animals, but they have some liberty of action in respect of burning vegetation, using fertilizers, and replacing hedgerows by wire fences. They can practice monoculture or over-graze. Where it becomes difficult to establish injury or fault, it seems that the best policy for preserving the amenity value of the countryside lies in the steady ventilating of the public interest. That interest includes not only the preserving of recreation areas and other contributions which the countryside makes to environment, but also the adaptation of western culture to a growing independence of the soil and the elements for food.

Conclusion

Outdoor farming is gradually being supplemented as well as being technically modified. Everyone knows about meat substitutes and the substances that promote health, fertility and growth. These are not merely byproducts of the chemist's skills, but part of the foundations of a new technology, just as fertilizers and artificial insemination were.

In such a complex industry as agriculture with its numerous products and production systems, it would be foolish to attempt a forecast of the likely shape of things to come, except in the broadest of terms. A substantially greater degree of self-sufficiency in Britain in sugar, cereals and meat —if that is a political objective—must be achieved by further intensification, since the reclaiming of enough arable land from presently unsuitable areas will be too expensive, or slow, or both. What form this intensification will take must vary, firstly, with the level already achieved and, secondly, with the finance available from other sources to purchase food for man and livestock. In Britain, the exploitation of the soil will be pushed further, and the agricultural landscape will lose some of its present charm as the pressure for grain increases at the expense of parks, trees, hedges and grassland. Meat and milk consumption will decline to amounts that can be produced on the remaining grassland, together with the hills, or can be afforded on the world market. Since they are derived from cereals, pig and poultry products are likely to be expensive, but the technology of indoor production will owe little if anything to traditional farm practices. Finally, the relationship between town and country will be changed, in such a way that the amenity value of the farmland will (with exceptions) be better appreciated, and wastage or damage severely discouraged. The old mutual dependence of urban and rural folk has been replaced by disinterest, and even intolerance, as the successes of the scientific revolution drove them apart; yet the time may not be far off when the grim realities of a Malthusian day of reckoning bring them together again.

If the neo-Malthusians are right this time, the world is indeed moving inexorably towards starvation on an increasing scale. If they are wrong, the increased food made available will be due to improved technology, better distribution, and economic encouragement of the growing of the necessary kinds of cereals. In either case, the problems for decision-makers will be how to maximize the output of human food for a balanced diet at a bearable cost, and to a much lesser extent how animals or people with special interests feel about methods of production. Because large areas of the world are unsuitable in the short term, say 20 years, for growing food for human consumption, but capable of supporting sheep, goats and cattle that can digest coarse fodder, these ruminants are likely to increase in relative importance. Human nature being what it is, pigs and poultry will not disappear, but they will survive in lesser numbers to feed the powerful. They will consume byproducts of milling and household waste, and we will make the most of their high efficiency as converters of their food to bacon and eggs.

The periodic famines that have marked the course of agriculture and the history of man still recur. Preventing or relieving them, however, poses such severe difficulties that freedom from hunger for all must remain a long-term objective. Agricultural technologists can be expected to master the biological problems gradually, while others struggle with the forbidding economic, social and logistic problems. Solutions which demand adaptation to change may be impeded, not only by the complexity of food production, but also because political institutions were not designed with philanthropy and self-sacrifice in mind. Already some countries (including Britain) where meat is an important item of diet, have reacted unfavourably to the idea of reducing the amount eaten in order to release grain for hungry people elsewhere. That these countries will sacrifice a significant part of their fertilizer supply in order to transfer some to the 60% of the world's population that now use only 15% of it seems unlikely, but they will piously recommend the growing of more food wherever possible.

In the short run, such advances as are made must come from knowledge and experience already available. It means more capital, more skill, more science. It means further stages of development in the direction of factory food production; and it means trying to release man from the uncertainties in his food supplies. Whilst disease, drought and infertility of the soil still threaten his crops and his livestock, it is much too soon to be patronizing about Nature. There are serious biological problems to solve as well as organizational and moral ones. Soil erosion due to over-grazing still occurs in spite of thousands of years of human experience of it. Forests are destroyed. Waste of food and energy still occurs on a large scale, and bacteria, fungi and insects become resistant to chemical methods of control.

For those who believe that man does not live by bread alone, there is an added dimension to this already intricate problem. To them, the mathematical model-making of scientists, economists and doomsters is an adolescent phase to be endured until the subjective values that spring from aesthetic and spiritual insights are learned. Not for them the regimentation which, as Dr Arnold Toynbee points out, is necessary if technology is to function properly.

The unexplained complexities of farms and farming include the varied reactions of way-of-life farmers, their attachment to their land and the conflicting interests of town and country. They preclude all generalizations except those of the coarsest approximation. Confronted by such an impenetrable thicket, it is not surprising that agricultural scientists concentrate their efforts where they are less likely to be frustrated—in pure technology. They are not working, and cannot work, for the ideal rural society where beauty, social justice and peace are paramount, but for the only structure they know which will deliver food and sometimes a profit. For the present they must rely on politicians to deal with the babel of voices from farmers, economists, environmentalists and the starving, but some attempt to harmonize them will eventually be made. It is perhaps not too ambitious to hope for some success in reconciling the objective with the subjective, and the rhetoric with the purpose. This will begin when society as a whole, and not just the special-interest groups, balances its need for food against its other uses for agricultural land.

FURTHER READING

Blount, W. P. (ed.) (1968), *Intensive Livestock Farming*, Heinemann, London.

Brambell, F. W. R. (1965), 'Rep. Tech. Comm. to enquire into the Welfare of Animals kept under Intensive Livestock Husbandry Systems', Cmnd. 2836, HMSO, London.

Carson, R. (1963), *Silent Spring*, H. Hamilton, London.

Ernle, Lord (1961), *English Farming, Past and Present* (6th ed.), Heinemann Educ. Books, London.

Fussell, G. E. (1972), *The Classical Tradition in West European Farming*, David and Charles, Newton Abbot.

Harrison, R. (1964), *Animal Machines*, Robinson and Watkins, London.

Jasiorowski, H. A. (1973), *Intensive Systems of Animal Production*, Third World Conf. Anim. Prod. Melbourne, Vol. 3.

Korach, M. (1964), 'The Science of Industry', in *Science of Science* (eds. Goldsmith, M. and Mackay, A.), Souvenir Press, London.

O.E.C.D. (1971), *Science, Growth and Society*, HMSO.

O.E.C.D. (1974), *Agricultural Policy in the United Kingdom*, HMSO.

Paddock, W. and P. (1968), *Famine—1975!*, Weidenfeld and Nicolson, London.

Toynbee, A. J. (1971), *Surviving the Future*, O.U.P.

Trevelyan, G. M. (1946), *English Social History* (3rd ed.), Longmans Green, London.

Index